Monographs of the Palaeontographical Society

The Palaeontographical Society was established in 1847, and is the oldest Society devoted to study of palaeontology worldwide. Its primary role is to promote the description and illustration of the British fossil flora and fauna, via publication of an authoritative monograph series. These monographs cover a wide range of taxonomic groups, from microfossils, trilobites and ammonites through to Coal Measure plants, mammals and reptiles, and from all ages from Cambrian to Pleistocene. They form a benchmark for understanding the past life of the British Isles and many include the original descriptions of numerous key species. The first monograph (on the Crag Mollusca) was published in March 1848 and the Society still continues this work today. Notable authors in the series include Charles Darwin (fossil barnacles) and Richard Owen (dinosaurs and other extinct reptiles). Beginning in 2014, the Cambridge Library Collection and the Society are collaborating to reissue the earlier publications, focusing on monographs completed between 1848 and 1918.

A Monograph on the British Fossil Echinodermata of the Oolitic Formations

Urged by his colleague Edward Forbes, Thomas Wright (1809–84) devoted himself to completing this monograph of the echinoderms ('spiny-skinned animals') of Britain's Oolitic Formations. These would be referred to as Middle Jurassic by the modern geologist. This is a notable contribution, describing as it does the echinoderms following a major stratigraphic gap. In the British Isles, apart from some minor occurrences in the Permian and Lower Jurassic, echinoderms are almost entirely absent from the Lower Carboniferous (Mississippian), a span we now know to represent 150 million years. Although common and diverse elsewhere during this interval, the British Oolitic echinoderms show many changes from those of the Mississippian. Wright's two-volume monograph includes thorough descriptions and locality details, all supported by beautiful plates. Volume 2, originally published in three parts between 1863 and 1880, considers those most beautiful of invertebrates, the asteroids (starfishes) and ophiuroids (brittle stars) of the Middle Jurassic.

Cambridge University Press has long been a pioneer in the reissuing of out-of-print titles from its own backlist, producing digital reprints of books that are still sought after by scholars and students but could not be reprinted economically using traditional technology. The Cambridge Library Collection extends this activity to a wider range of books which are still of importance to researchers and professionals, either for the source material they contain, or as landmarks in the history of their academic discipline.

Drawing from the world-renowned collections in the Cambridge University Library and other partner libraries, and guided by the advice of experts in each subject area, Cambridge University Press is using state-of-the-art scanning machines in its own Printing House to capture the content of each book selected for inclusion. The files are processed to give a consistently clear, crisp image, and the books finished to the high quality standard for which the Press is recognised around the world. The latest print-on-demand technology ensures that the books will remain available indefinitely, and that orders for single or multiple copies can quickly be supplied.

The Cambridge Library Collection brings back to life books of enduring scholarly value (including out-of-copyright works originally issued by other publishers) across a wide range of disciplines in the humanities and social sciences and in science and technology.

A Monograph on the British Fossil Echinodermata of the Oolitic Formations

VOLUME 2

THOMAS WRIGHT

CAMBRIDGE
UNIVERSITY PRESS

CAMBRIDGE
UNIVERSITY PRESS

University Printing House, Cambridge, CB2 8BS, United Kingdom

Cambridge University Press is part of the University of Cambridge.

It furthers the University's mission by disseminating knowledge in the pursuit of
education, learning and research at the highest international levels of excellence.

www.cambridge.org
Information on this title: www.cambridge.org/9781108081160

This edition first published 1863–80
This digitally printed version 2015

ISBN 978-1-108-08116-0 Paperback

THE

PALÆONTOGRAPHICAL SOCIETY.

INSTITUTED MDCCCXLVII.

VOLUME FOR 1881.

LONDON:

MDCCCLXXXI.

MONOGRAPH

ON THE

BRITISH FOSSIL

ECHINODERMATA

OF

THE OOLITIC FORMATIONS.

BY

THOMAS WRIGHT, M.D., F.R.S., F.G.S.,

VICE-PRESIDENT OF THE PALÆONTOGRAPHICAL SOCIETY; CORRESPONDING MEMBER OF THE ROYAL SOCIETY OF SCIENCES
OF LIÈGE; THE SOCIETY OF NATURAL SCIENCES OF NEUCHÂTEL; VICE-PRESIDENT OF THE COTTESWOLD
NATURALISTS' FIELD CLUB; CONSULTING SURGEON TO THE CHELTENHAM HOSPITAL;
AND MEDICAL OFFICER OF HEALTH TO THE URBAN SANITARY DISTRICTS
OF CHELTENHAM, CHARLTON KINGS, AND LECKHAMPTON.

VOLUME II.—THE ASTEROIDEA AND OPHIUROIDEA.

LONDON:

PRINTED FOR THE PALÆONTOGRAPHICAL SOCIETY.

1863—1880.

J. E. ADLARD, BARTHOLOMEW CLOSE.

PRINTED BY
J. E. ADLARD, BARTHOLOMEW CLOSE.

INTRODUCTION.

In bringing my Monograph on the Asteroidea and Ophiuroidea to a close a few words of explanation appear to be necessary in order to avoid any misconception as to the cause that has occasioned delay in the completion of the work.

When I had assembled for the first time all the materials I had collected for the volume I found, much to my regret, that the specimens were fewer and more fragmentary than I anticipated, and that it was impossible to carry out the description of the families in a manner similar to the one I had adopted in the Echinoidea. I therefore determined to figure and describe all the species that I had collected and wait for the discovery of others in public and private collections which I had not at that period been able to inspect. The part containing the Asteroidea appeared in 1863, and the part on the Ophiuroidea in 1866. Since then I have made diligent search among all collections that were likely to contain Jurassic Echinoderms, and after many a hunt I have only succeeded in obtaining two additional species from these sources, and finding a much better specimen of an important Yorkshire species, *Astropecten rectus*, of which I have given a good figure.

In the usual progress of discovery by waiting patiently for new things sometimes a few specimens are met with, and I am happy to say that my patience has been rewarded. Among the Goniasteridæ a very fine specimen of *Stellaster* was collected by my friend Samuel Sharp, Esq., F.G.S., from the ironstone beds of Inferior Oolite near Northampton, certainly one of the finest fossil Star-fishes which the Inferior Oolite has hitherto yielded up. An allied species was soon afterwards discovered in the " Calcaire à Entroques," a bed of the same age, at Mâcon (France).

Among the family URASTERIDÆ a small specimen belonging to the genus *Uraster* has been found in the Forest Marble of Wilts, of which I have given figures.

The OPHIUROIDEA have been quite as scarce as the true Star-fishes. The "*Avicula contorta beds,*" Upper Trias, at Hildesheim, were found to contain specimens of *Ophiolepis Damesii*, Wr., a species which I first described from specimens sent for my determination from the Berlin Museum by Dr. Dames. These notes were subsequently translated and inserted in 'Der Zeitschrift der Deutschen geologischen Gesellschaft,' Jahrgang, 1874, with an excellent figure of the species. A few months later the same Brittle-Star was collected from the black shales belonging to the Bone-bed series at Westbury-on-Severn, and soon afterwards similar black shales above the Bone-bed near Leicester yielded remains of the same species.

After my additional plate had been printed, and the last sheet of this volume been twice revised, I received, on the 3rd inst., from my friend Professor Buckman, F.G.S., of Bradford Abbas, for description in the 'Proceedings of the Dorset Naturalist Field Club,' a specimen which he had collected from the Calciferous Grit at Sandsfoot Castle, near Weymouth. This appears to me to be a new *Ophiurella* from our Corallian strata, and I have had it drawn on wood and inserted in its natural place in the text.

I have grouped the ASTEROIDEA into four families—I. URASTERIDÆ; II. TROPIDASTERIDÆ; III. GONIASTERIDÆ; and IV. ASTROPECTINIDÆ. Of these, URASTERIDÆ and TROPIDASTERIDÆ (not defined in the text) may receive the following diagnosis :—

The URASTERIDÆ, *Wright,* have a stellate, five-rayed body; the rays are round or angular and abundantly covered with spines. The ambulacral areas are lanceolate and bordered by several rows of spines, and the upper surface of the disk and rays provided with short, blunt, and thorn-like spines; in some species they are sparsely distributed in single rows, or in others closely set together in a linear arrangement on the disk rays. The interspinous tegumentary is naked and perforated with pores for respiration; there are four rows of pores for the passage of the tubular feet, so that the pores have a quadriserial arrangement in the avenues. The pedicellaria are supported upon soft stems, and the opening of the vent is dorsal and excentral. This family ranges from the Lias seas down into those of our own time, with so little variation in anatomical structure that the closest scrutiny only detects the slightest specific modification between the Urasters of the Lias and those from our shores.

The TROPIDASTERIDÆ, *Wright*, have a stellate body with short rays and variable as to numbers. The upper surface is covered with solitary or fasciculated spines, arranged in regular order. The ambulacral areas are bordered by fasciculi of spines disposed in rows more or less numerous. There are two rows of pores for the passage of the tubular feet, by which they are distinguished from the URASTERIDÆ, in which they are quadriserial. The vent is dorsal and excentral in the *Solasters*. This family ranges from the Lias seas to those of our day. The Tropidasters lived in the Lias period, and the *Solasters* have lived on from the Lower Jurassic times into the present time.

The ASTEROIDEA are very well represented by eight genera, and twenty-three species in English Jurassic strata, and the anatomy of these skeletons has been fairly made out in these native fossil forms.

The OPHUROIDEA are grouped into two families, the OPHIURIDÆ and ASTEROPHYDIÆ.

The OPHIURIDÆ contain five genera that have representatives in our Jurassic strata. The structure of the body-disk in this family is so delicate and fragile, and consequently more or less injured or utterly destroyed in the fossil state, that its structure is made out with great difficulty, and often with much uncertainty. Should better specimens be discovered hereafter with their anatomical characters better preserved, then our errors, if any, can be corrected. With the materials at my disposal I have been scrupulously careful, with the aid of the lens and the microscope, to submit all these parts to a most minute inspection; still we cannot revive traces of organic structure when they are hopelessly effaced, therefore some of my diagnoses of genera and species are neither as complete or precise as I should have wished them to be from causes which I was unable to control.

My most kind and considerate friend, our worthy secretary, the Rev. Thos. Wiltshire, F.G.S., knowing how much my time is occupied with public duties, has generously prepared a summary and analysis of the families and genera of the ECHINOIDEA, ASTEROIDEA, and OPHIUROIDEA, described in the two volumes of the ' Oolitic Echinodermata.' This important addition to my work will be very useful to students, as it brings together in a condensed form the heads of information dispersed through the two volumes. The nomenclature of the text has been preserved, and the range of the species in time carefully noted. My friend has likewise added—*first,* a tabular list of the genera, with their ranges in geological time; *secondly,* a list of the genera and species described in Volumes I and II; and *thirdly,* an Index to Volume II. I have no doubt these valuable additions to the practical worth of the volumes will be as much appreciated by my readers as they are by their author, who tenders his very best thanks and warmest acknowledgments to his old and much valued friend for this kind and unsolicited contribution to his work.

THOMAS WRIGHT, M.D., F.R.S.

4, St. Margaret's Terrace, Cheltenham;
15th *March*, 1880.

A MONOGRAPH

ON THE

BRITISH FOSSIL

ECHINODERMATA

FROM

THE OOLITIC FORMATIONS.

BY

THOMAS WRIGHT, M.D., F.R.S.E., F.G.S.,

CORRESPONDING MEMBER OF THE ROYAL SOCIETY OF SCIENCES OF LIEGE,
AND SENIOR SURGEON TO THE CHELTENHAM HOSPITAL.

VOLUME SECOND.

PART FIRST.
ON THE ASTEROIDEA.

LONDON:
PRINTED FOR THE PALÆONTOGRAPHICAL SOCIETY.
1862.

J. E. ADLARD, PRINTER, BARTHOLOMEW CLOSE.

A MONOGRAPH

ON THE

FOSSIL ECHINODERMATA

OF THE

OOLITIC FORMATIONS.

THE ASTEROIDEA.

THE true star-fishes forming the order ASTEROIDEA have in general a depressed stelliform body, which sometimes assumes a polygonal or pentagonal figure. From the central disc five or more hollow rays proceed, containing prolongations of the internal organs (Pl. I, fig. 1). The entire upper surface is covered with a coriaceous integument, in which a series of calcareous pieces, often supporting spines, tubercles, and pedicellariæ, are developed (fig. 2, *a*). In the centre of the under surface is the mouth-opening, from whence radiate to the extremities of the rays as many ambulacra as there are lobes; in these, the tubular retractile feet are arranged in two or four rows; and the margins of the rays in many genera are bordered by well-developed spinigerous plates. (Pl. I, fig. 2, *b*.)

The skeleton of the ASTEROIDEA is a very complicated framework. It is composed of a great number of little bones or ossicula, articulated together in such a manner as to combine strength with flexibility. The ossicles vary in form and number in different parts of the skeleton; they have a persistent arrangement in the various genera, so that the ossicula of a star-fish afford us good evidence of the rank of its owner among the radiata, as the bones of a reptile or a mammal do amongst the vertebrata. The comparative anatomy of the skeleton of the ASTEROIDEA has not yet been worked out in

1

many genera, but we recommend the study of Tiedemann,[1] Meckel,[2] Sharpey,[3] and Müller's[4] works on the anatomy of some common species, as examples of what may be achieved in other groups, if correct observation and like diligence be brought to the task.

As the skeleton of the ASTEROIDEA is that part of their bodies which is most frequently preserved in a fossil state, it is necessary that we should be well acquainted with the structure and relations of its different component parts. If, for example, we remove from the common star-fish, *Uraster rubens*, Lin., the integument and spines from the upper surface of the disc and rays, and afterwards the viscera enclosed therein, the structure of the ambulacral portion of the skeleton will be well exposed. It is seen to

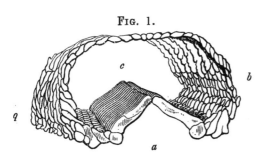

FIG. 1.

Section of a ray of *Uraster rubens* showing the arrangement of the calcareous ossicula.

consist of a central ring surrounding the mouth-opening, composed of ten larger and five smaller pieces firmly united together by ligaments; the ten larger pieces are disposed in pairs opposite the base of each ray, and the five smaller pieces occupy the interbrachial angles; the ten elements of the oral ring are perforated, for the passage of soft tubular organs.

Each ray is composed of a considerable number of small bones or ossicles, which form rings, as seen in fig. 1, representing the section of a ray of *Uraster rubens*, Lin. These bony circles succeed each other from the base to the apex, each segment being a repetition in form and position of all the others; the size of the rings, however, diminishing gradually from the base to the apex. The ossicles at the under part of the ray (*a*) are symmetrical, and articulated together in such a manner as to permit of considerable motion; their upper surface forms the floor of the cavity, in which prolongations of the digestive and other vital organs are contained (*c*); the under surface of the ossicles forms the ambulacral valley through which the tubular suckers pass. In fig. 1 *a* the two long femur-like bones at the bottom of the ring project obliquely upwards and inwards, and join each other in the median line; they are articulated at the base with other ossicula, which I shall presently describe. To the lateral parts of this central framework another series (*b b*) of larger ossicles are joined, which rise nearly parallel to each other like ribs encircling a thorax; they are connected by transverse osseous bars, and the whole is enveloped in the tegumentary membrane which encloses the upper portion of the rays. The ossicula are lined internally by a white, tough, fibrous membrane, which extends to the sides and floor of the ray, unites the ossicles together, and contributes to form

[1] Tiedemann, 'Anatomie der Röhrenholothurie des pomeranzenfarbigen Seesterns und Steinigels, Landshut, 1816, folio.

[2] Meckel, ' System der vegleichenden Anatomie,' Band ii, p. 19.

[3] 'Cyclopædia of Anatomy and Physiology,' Art. Echinodermata, from which figs. 1 and 2 are copied.

[4] Joh. Müller über den Bau der Echinodermen. 1854. 4to.

the compact, flexible mechanism so admirably exhibited in the arms of many species. Fig. 2 represents the lateral view of part of a ray of *Uraster rubens*, Lin., and exhibits the net-like arrangement of the ossicles on the sides and upper surface; this reticulate structure forms the inter-ambulacral portion of the skeleton.

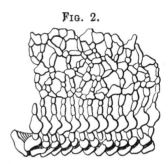

FIG. 2.

Reticulate arrangement of the ossicles on the sides and upper surface of a ray in *Uraster rubens*.

The test of the ECHINOIDEA is formed of ambulacral and inter-ambulacral areas, and a similar arrangement of the ossicula may be observed in the skeleton of the ASTEROIDEA; the centrum of the disc or oral ring in the ASTERIADÆ is the homologue of the centrum and auricular arches in the test of the ECHINIDÆ, to which the muscles of the jaws are attached; the centrum in both orders therefore forms the arch to which the ossicles of the rays in the star-fishes, and on which the plates in the sea-urchins, are supported.

The arches forming the central part of the base of the rays are the homologues of the ambulacral areas in the ECHINIDÆ; they are composed of two central, oblong principal pieces (fig. 1 *a*) united at the median line, and two smaller transverse pieces on which they rest, having two smaller inferior pieces external to the preceding; these six elements enter into the composition of a single segment of the ambulacral area. In a specimen of *Uraster rubens*, prepared as already described, for the purpose of displaying the skeleton, I have counted 140 ambulacral arches in each ray, which multiplied by 6 for the six elements in each arch, $140 \times 6 = 840$ ossicles in one ray; this multiplied by 5 for the five rays, $840 \times 5 = 4200$, is the number of ossicles in the ambulacral portions of the skeleton of this specimen, exclusive of the elements of the centrum. The ossicles forming the lateral and upper portions of the ray are the homologues of the inter-ambulacral areas in the ECHINOIDEA (fig. 2). The number of separate pieces entering into the composition of this part of the skeleton is very great, arising from the smallness of the bones, and the diverse forms of their reticulate arrangement in the different genera, to form a structure at once resistant and flexible, and adapted to the habits of the organism (fig. 2); the inter-ambulacral areas of the ASTERIADÆ are for this reason very unlike the homologous portion of the test in the ECHINIDÆ, where these areas consist of two columns of broad spinigerous plates, between which the narrow ambulacra are placed. In the ASTERIADÆ, on the contrary, the ossicular elements of the inter-ambulacra, besides their locomotive functions, have assigned to them the formation of the sides and roof of the hollow cylindrical arms (fig. 1 *c*).

The structure of the rays varies so much in the different genera of this order, that any general description would necessitate the enumeration of so many exceptions to the common plan of organization, that I prefer pointing out the differences which several of the genera exhibit, rather than attempt to give a general outline of the entire group.

In the genus *Uraster* (Pl. I, fig. 2 *a, b*) the rays are long, and the ossicula on each side of the ambulacral valley support many rows of spines; the ossicles on the sides and upper surface of the ray form a hollow cavity for lodging the viscera (fig. 1 *c*); and the numerous small bones entering into this net-like structure support blunt or pointed spines (fig. 2 *a*), the integument between the osseous pieces is naked, and perforated by pores which communicate with the interior. Numerous pincers-like pedicellariæ, supported on soft stems, encircle the basis of the spines, or are distributed amongst them, whilst others are disposed at the angles of the rays. All the *Urasters* possess an excentral anal opening.

In Plate I, fig. 2 *a*, I have figured a portion of the upper surface of a ray of *Uraster tenuispinus*, M. and T., which shows the recurved spines raised upon the inter-ambulacral ossicles, and the naked integument between the spines perforated with respiratory pores. Pl. I, fig. 2 *b*, is the under surface of the same ray; four rows of tubular sucking-feet occupy the ambulacral valley, which is fringed with two rows of small spines arranged in an oblique comb-like order; external to these other rows of larger spines arm the lateral parts of the ray. Pl. I, fig. 3, represents a portion of the ambulacral skeleton, and shows the spaces for the passage of the tubular retractile feet.

The *Astropectens* have a stellate body, flattened on both sides, and furnished with two rows of tubular feet; in the ambulacral valleys, the rays are narrow and elongated, and their sides are bordered with two rows of large, regular, marginal plates (fig. 3). The ventral marginal plates carry long moveable spines, and the dorsal series in general are covered with granules more or less developed, which sometimes likewise carry spines. On the upper surface of the ray the inter-marginal space is thickly set with bunches of *paxillæ*, and the under surface is crowded with regular rows of short spines.

All these parts are well seen *in situ* in the section of a ray of the common Butthorn (*Astropecten aurantiacus*, Lin.) If the ray is laid open along the middle of the upper surface, and the dorsal integument and marginal plates are folded over into a horizontal position, the homology of the different parts of the ray, with the ambulacra and inter-ambulacra of the *Echinidæ* becomes evident.

The annexed figures of the upper and under surfaces of a ray of *Astropecten polyacanthus*, M. and T., shows the general structure of this part in the genus *Astropecten*. The ambulacral areas have two rows of tubular retractile feet; the valley is bounded by two rows of short, stout spines, arranged in groups by the side of the suckers, each of the lower marginal plates is armed with four long, recurved, tooth-like spines (fig. 3 *A*), and other shorter spines cover the plates at their base; the upper marginal ossicles likewise support long, stout,

FIG. 3.

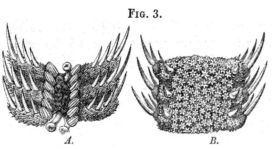

A.   B.

Portion of a ray of *Astropecten polyacanthus*, M. and T.   *A*, under surface; *B*, the upper surface of the ray.

recurved spines, and the entire upper surface of the ray is crowded with stellate paxillæ, (fig. 3 *B*.)

In the *Solasters*, the disc is large; the rays, from twelve to fifteen in number, are short, about half the length of the diameter of the body. In the common Sun-star, *Solaster papposa*, Lin., one of the most common and handsome of British star-fishes, the structure is well exposed; fig. 4 *B* exhibits the upper surface, and fig. 4 *A* the under surface of this species. The disc is large and flat, one half the diameter of the body; the rays, in general twelve in number, are round and short, their length being about one half the diameter of the disc.

FIG. 4.

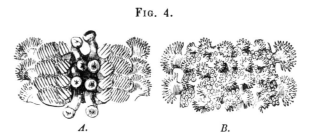

A.   *B.*

Portion of a ray of *Solaster papposa*, Linn. *A*, the under; *B*, the upper surface.

The entire upper surface is covered with tubercles, those on the disc are scattered over its surface, whilst they are arranged in regular rows on the rays. Each of these tubercles bears a bundle of long, spiculiform, striated spines, fig. 4 *B* having from eighteen to twenty grouped in each fasciculus. In the rays there are in general five or six rows of spiniferous tubercles, those on the borders being the largest. The integument between the tubercles is naked, and perforated with many tentacule-pores (fig. 4 *B*); there are no pedicellariæ, and the vent is central; the madreporiform plate is excentral, and its surface is covered with fine radiating lamellæ. The under surface of the rays (fig. 4 *A*) are narrowly lanceolate, the avenues have two rows of suckers, and the ambulacral plates support longitudinal bundles of spines, four or five in each fasciculus. External to these are regular, transverse rows of spines, supported on transverse ridges, eight or ten in each row. "The third series forms a bordering to the arms, and consists of sets of from eighteen to twenty long, fasciculated spines, placed on broad, compressed, articulated bases. The mouth is protected by a beautiful and peculiar mechanism. The angles formed by the joined origins of the rays each bear an ovate sub-triangular plate, grooved down the centre, and carrying two semicircles of long tapering spines, which project in a comb-like manner over the mouth."[1]

In the structure of the rays, the genus *Pteraster* resembles some remarkable fossil species. The tegumentary membrane on the convex upper surface is furnished with rows of short spines (fig. 5 *B*); the under surface has a biserial arrangement of tubular suckers in the ambulacra, and their margins are provided with numerous transverse fasciculi of spines, five or six in number, the spines of each fasciculus are connected together by a thin membrane, and

[1] Forbes, 'British Star-fishes,' p. 113.

form a series of fan-like structures (fig. 5 *A*) ; the borders of the rays are likewise armed with long spines, extended transversely from the sides. These spines are connected by a fold of tegument, and united together like the ray-bones in the fin of a fish. Fig. 5, from Müller and Troschel,[1] who described this genus, from the North Sea, shows the regular disposition of these fan-like fasciculi of spines at the borders of the ambulacral valley, and the long transverse

FIG. 5.

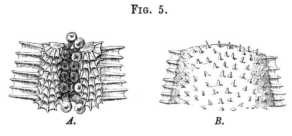

A.   *B.*

A portion of a ray of *Pteraster militaris*, M. and T.
A, the under ; *B*, the upper surface.

spines at the margin of the rays, both classes of spines being in this genus connected together by prolongations of the common tegumentary membrane.

In *Luidia*, the body is stellate ; the rays are long, flat, and narrow, with a single row of ventral marginal plates supporting long spines ; the upper surface of the ray is closely set with paxillæ (fig. 6 *A*) ; the ambulacral valleys are narrow (fig. 6 *B*), the suckers biserial, and two sets of spines occupy the under side of the ray. The long, recurved spines on the margin of the rays, with the paxillæ covering their upper surface, ally this genus

FIG. 6.

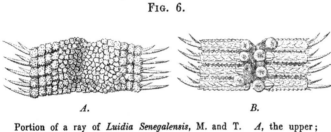

A.   *B.*

Portion of a ray of *Luidia Senegalensis*, M. and T.   *A*, the upper ;
B, the under surface.

to *Astropecten*. The toughness of the body and arms, in some star-fishes, is not more remarkable than their fragility in others ; and the difficulty attending the capturing of an entire specimen of *Luidia*, from its voluntary destructiveness, has been so graphically recorded by my lamented colleague, that I cannot do better than quote his account. "It is the wonderful power which *Luidia* possesses, not merely of casting away its arms entire, but of breaking them voluntarily into little pieces with great rapidity, which approximates it to the *Ophiuræ*. This faculty renders the preservation of a perfect specimen a very difficult matter. The first time I ever took one of these creatures, I succeeded in getting it into the boat entire. Never having seen one before, and quite unconscious of its suicidal powers, I spread it out on a rowing-bench, the better to admire its form and colours. On attempting to remove it for preservation, to my horror and disappointment, I found only an assemblage of rejected members. My conservative endeavours were all neutralised by its destructive exertions, and it is now badly represented in my cabinet by an armless disc and a discless arm. Next time I went to dredge on the same spot,

[1] Müller and Troschel, 'System der Asteriden,' p. 128.

determined not to be cheated out of a specimen in such a way a second time, I brought with me a bucket of cold fresh water, to which article star-fishes have a great antipathy. As I expected, a *Luidia* came up in the dredge,—a most gorgeous specimen. As it does not generally break up before it is raised above the surface of the sea, cautiously and anxiously I sunk my bucket to a level with the dredge's mouth, and proceeded in the most gentle manner to introduce *Luidia* to the purer element. Whether the cold air was too much for him, or the sight of the bucket too terrific, I know not; but in a moment he proceeded to dissolve his corporation, and at every mesh of the dredge his fragments were seen escaping. In despair, I grasped at the largest, and brought up the extremity of an arm, with its terminating eye, the spinous eyelid of which opened and closed with something exceedingly like a wink of derision."[1]

The *Goniasteridæ* have pentagonal bodies, flattened on both sides; the margin is bounded by two rows of large marginal plates, larger than those on other parts of the disc, and both entering into the formation of the border (figs. 7 *c*); their surface is variously covered with granules, spines, or pedicellariæ, and they are often encircled by granules. The upper surface of the disc and rays, within the marginal plates, is composed of small, flat, hexagonal, pentagonal, or tetragonal ossicula (fig. 8 *A*), and a like armature covers the under surface; the ambulacral avenues are bordered by a series of square ossicula, which are often marked with parallel grooves for lodging the spines (fig. 8 *B*). Towards the extremities of the rays (fig. 7 *d*), the dorsal border-plates are variously modified for lodging and protecting the eyes. Fig. 7, after Müller, represents *Astrogonium cuspidatum*, M. and T., laid open from above to show, *a*,

FIG. 7.

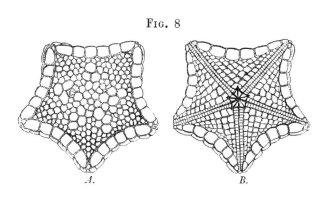

Astrogonium cuspidatum, M. and T.

the ambulacral plates; *b*, the inner surface of the inter-ambulacral plates; *c*, the upper border-plate; and *d*, the terminal plate, modified to protect the eye.

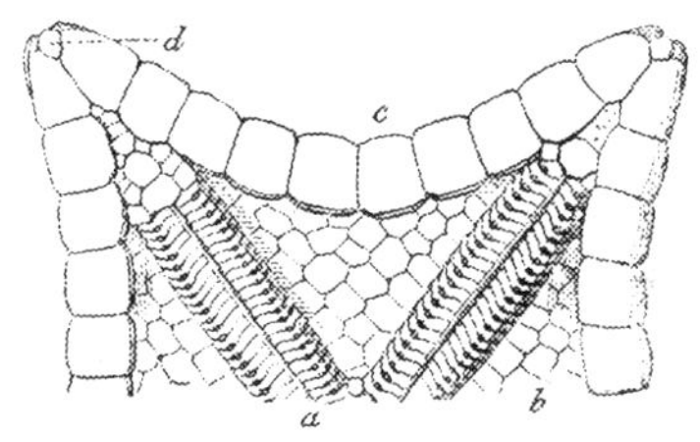

Fig. 8 shows the upper and under surfaces of a small *Astrogonium* in the British Museum Collection; *A* is the upper surface, exhibiting the large superior border-plates, enclosing the small polygonal discal plates with their granular circles, which occupy the whole intra-marginal upper surface; *B* shows the base, with the large inferior border-plates, and the

FIG. 8

A. *B.*

[1] Forbes, 'British Star-fishes,' p. 138.

tesselated inter-ambulacral plates forming the floor of the rays, with the narrow ambulacral ossicles and the stellate mouth-opening.

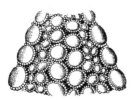

FIG. 9.

Upper surface of a ray of
Astrogonium magnificum, M. and T.

In *Astrogonium magnificum*, M. and T., the superior border-plates are encircled two thirds by rows of granules; and the upper surface of the arm is covered with large circular or oblong plates, smooth and convex on their upper surface, and each surrounded by a complete circle of granules. This structure is exhibited in the annexed figure 9, from Müller and Troschel.

In the genus *Stellaster*, the upper and under sides of the pentagonal body are flat, and

FIG. 10.

Under surface of a ray of *Stellaster*
Childreni, Gray.

surrounded by two rows of large marginal plates, both of which enter into the formation of the high border; each of the lower marginal plates carries, near the outer side, a flat, moveable spine (fig. 10); and several granules are scattered over the surface. The ambulacra are narrow, and the suckers biserial; both sides of the intra-marginal disc are covered with granulated plates, on which numerous pedicellariæ are fixed. Fig. 10, which exhibits the under surface of one of the rays in *Stellaster Childreni*, Gray, illustrates the characters of this genus.

In *Ophidiaster* the rays are long, cylindrical, or conical; the osseous framework of each consists of a series of ossicula of two or three different forms; in this section of a ray (fig. 11) there are seven rhomboidal ossicula, of which three belong to the upper surface,

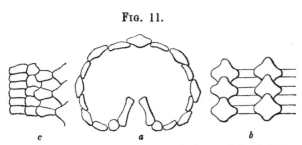

FIG. 11.

c   a   b
Section of a ray of *Ophidiaster*, with the inter-ambulacral plates.

and two to each of the sides; eight oblong ossicula (*a*) unite the rhomboidal pieces (*b*) together, two uniting the rhomboidal ossicles of the upper surface of the ray with each other, two linking the superior lateral with the surface plates, two connecting the pairs of lateral rhombs of each side (*b*), and two articulating the inferior laterals with the small, round ossicles which link the inter-ambulacral plates (*a*) with the long femur-like ambulacral ossicula projecting upwards into the interior of the arm (fig. 11 *a*). In this transverse section of a ray of an *Ophidiaster* we find nineteen ossicula, of which four or six belong to the ambulacral area, and the others to the inter-ambulacral portion. If the ossicula of an arm, therefore, were folded down, and extended outwards, the six ambulacra would form the centre, and the other plates on each side would represent the two halves of the adjoining inter-ambulacral areas.

In *Arthraster* the number of the ossicula is less, and their arrangement different, than in *Ophidiaster*. This genus, which is cretaceous and extinct, has, according to Forbes, only seven ossicula in the framework of the ray, as shown in the transverse section, fig. 12 *a*, exclusive of the ambulacral bones, which are unknown. These plates are articulated together in such a manner as to form a compact armature surrounding the arms (fig. 12 *a*), like the plates and sutures in the test of *Echinidæ*, of which they are the homologues. All the seven ossicles are

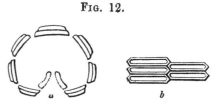

FIG. 12.

Section of a ray of *Arthraster*, with the side plates.

similar in form, each consisting of a transversely oblong, expanded, linear base, terminating in an acute angle at each end (fig. 12 *b*), and bearing along the centre a crest-like ridge with steep sides. The central ossicle is the largest, and this Forbes regarded as the equivalent of all the ossicles in the upper surface of the arm of *Ophidiaster* (fig. 11), whilst the others may be considered as the homologues of the lateral and ventral plates, with their connecting ossicula.[1]

The genus *Oreaster* comprises a group of pentagonal star-fishes, which have the under surface flat, and the upper surface more or less elevated; large tubercles or globular calcareous spines occupy various parts of the dorsal surface (fig. 13); the skeleton is formed of large plates, mostly of an irregular polygonal shape, which are disposed on the ridges of the arms in a more or less squamated order; the margins of the rays are surrounded by two rows of granulated plates, which overlap each other, the dorsal border-plate alone forming the margin, and the ventral border-plate lying on the under surface. There are two rows of tubular feet in each avenue, and a subcentral vent on the dorsal surface, which is elevated and sub-pyramidal (fig. 14). The interior of the test is strengthened by calcareous pillars, as in the *Clypeasters* among the ECHINIDÆ. The valve-like pedicellariæ are all sessile.

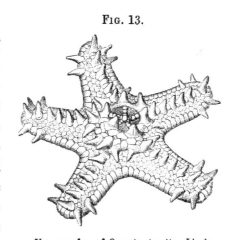

FIG. 13.

Upper surface of *Oreaster turritus*, Linck.

[1] 'Memoirs of the Geological Survey,' vol. ii, part 2, p. 467. See also Dixon's 'Geology and Fossils of Sussex,' p. 336, pl. 23, fig. 1.

On the Homology of the Skeleton in the Asteriadæ.

Although the homology of the skeleton of the *Asteriadæ* has long engaged the attention of naturalists, still upon this subject much diversity of opinion prevails, scarcely two of the classical authors holding the same views as to the relation existing between the test of a sea-urchin and the skeleton of a star-fish. Under these circumstances, I purpose giving copious extracts from the works of Delle Chiaje, De Blainville, Müller and Troschel, Agassiz, Müller, and Huxley, with the view of placing this interesting subject fairly before the reader.

Osseous System.

The inferior part of the rays in the *Asteriæ* (says Delle Chiaje),[1] or the whole of the rays in the *Ophiuræ*, is composed of a series of fragmentary and semicircular bones, almost similar to the vertebræ, the disposition of which deserves a special study (Cuvier, ' Rég. Anim.,' tom. iv, p. 9). The bones placed around the mouth are five in number; each of these is composed of four articulated parts; that is, two at the upper end, connected by useful teeth and corresponding ligaments, rounded at the bottom, and spinous; each of them, besides, is composed of as many (four) cylindrical, lateral bones, joined with the branches of the other four great vertebræ.

Thus for each ray there is a series quite decreasing, and each of them is made of two denticulated pieces, provided with ligaments, which have a hole underneath for the passage of the vertebral artery, and besides, of two faces connected with another spinous egg-shaped piece, which shuts the aperture in each ray; to these are fastened the feet, and in its inside are sometimes found two small annelides, one of which appears to me to be the same as the one described, although roughly, by the illustrious Baster (Opusc. subsec., tom. iv, fig. 9).

Some other imbricated spines, more or less short, are turned towards the sides of the ray, which at the lateral undermost part terminates in a long articulated spine, near which is found the hole for the passage of sea water, and by another smaller one lodged within the apex. Between this and the vertebra a long piece, according to the amplitude of the ray, articulates itself transversely. As in *Astropecten arantiacus*, the vertebræ are sufficiently large, so the ampullæ of the radial arteries fill up the entire space; whilst in *Echinaster echinophora* the vertebræ being smaller, the ampullæ fill up the space alternately. The uppermost portion of the ray is found likewise to be formed of a chain of osseous pieces, which are sometimes long and sometimes short. The same conformation is observable in *Uraster rubens*.

Besides the file of vertebræ in the rays of *Asterias exigua*, we notice between each of them many small cuneiform, imbricated bones, formed in the shape of as many triangles

---

[1] Memorie sulla storia e notomia Degli Animali senza Vertebre, vol. ii, p. 289. I am indebted to my friend, M. Ronna, for this translation from Delle Chiaje.

as there are spaces in each ray : in the angle at the vertex of this the osseous column is fastened to the superior integument, which appears to be perforated. The texture of *Asterias rosacea* is likewise entirely osseous.

The rays in the *Ophiuræ* have compressed vertebræ, orbicular, without any perforation, and with articular faces, as well as two furrows, one upwards and the other downwards. In the neighbourhood of the mouth, where they get larger, they support the two branches, of which the denticulated jaw is composed towards its end, and in *A. cordifera* (*Ophiolepis ciliata*) towards its basis. On the sides of the radii in *A. ophiura* (*Ophiolepis scolopendrica*) is noticed a couple of lamellated bones which are connected with the radii and with the epidermis : in case of this epidermis being deficient, as in *A. cordifera* (*Ophiolepis ciliata*), these scaly bones are connected with its osseous and imbricated crust.

Echinaster echinophora has, moreover, many small bones, still smaller in *Uraster rubens*, which are articulated with the very small bones constituting the superior surface of the body. They correspond likewise with the axis of the moveable tubercles, acuminated in *E. echinophora*, rounded in *Asterias Savaresi*, and surrounded by the epidermis ; from these shoot out several muscular fibres, directed towards the respective osseous pedicellariæ, which, when looked upon through a magnifying lens, appear either to have an acuminated shape, or to be compressed and entirely rounded like the bill of a goose. Each pedicellaria is composed of two osseous, articulated pieces, fixed on a common basis of the same description. They enjoy the faculty of adhering to adjacent bodies, and keeping closely adhesive.

The small osseous chalices (or paxillæ) of *Astropecten arantiacus, A. bispinosa,* &c., are differently constituted. Each of these is a cylinder fastened at the bottom by means of strong muscular bands, the fibres of which have several intervening holes ; it ends at the top in a convex shape by many cylindrical pieces, distributed according to a double series, with internal articulations, and provided in the middle with a conical piece, excepting in *A. arantiacus* alone. It would be useless to enter into further minute details, which may be more easily traced out by an inspection of the purposely drawn-out figures in the tables.

De Blainville[1] says, the star-fishes have still a particular disposition of the external envelope, the dermis is more distinct than in the urchins ; we see better that solid and calcareous parts are developed in its interior. These parts form spines or scales more or less immoveable, and which present dispositions proper to each of the groups of this order.

In the *Asterias,* properly so called, that is to say, in the species in which the body is not provided with appendages, but which are divided more or less deeply into rays hollowed out inferiorly by a groove which extends throughout their entire length ; the superior parts have the skin sometimes soft, and oftener solidified by a greater or less number of irregular pieces arranged in a reticulate manner.

They are sometimes almost smooth, but in general are bristled with tubercles of

[1] De Blainville, 'Organisation des Animaux,' p. 213.

different sizes, disposed more or less irregularly, and which furnish excellent specific characters. The lateral and inferior parts are, on the contrary, sustained by a greater number of pieces, much more regularly disposed. They acquire sometimes a very great development, as in the *Asterias tessellata*, Lamk.; they appear to me always to form three series, one superior, the second altogether lateral, and the other inferior. It is these which unite with the series of pieces I call ambulacral, because it is between them that the tubular feet escape, as in the sea-urchins. The last two series of lateral pieces carry moveable spines; they are still the analogues of the inter-ambulacral areas of the sea-urchins; these spines vary, nevertheless, both as regards their figure and the number of rows which they form; they are always very singular for their resemblance to a grain of corn. As to the ambulacral pieces, they are very regular, very symmetrical, and they resemble in the median line of the inferior part of each ray a kind of spine, which sustains it, and which permits movements between its numerous articulations, as in a species of vertebral column.

Müller and Troschel,[1] in their 'System der Asteriden,' state, that the *Asteriadæ* are Echinodermata of a stellate, or polygonal, mostly of a pentagonal form. In addition to the tegumentary skeleton, they possess an internal skeleton, which is wanting in all the others. This consists of as many rows of pieces moveably articulated together, as there are lobes in the body, and which always proceed from the circumference of the mouth, and from the under side of the rays. In the *Asterias* these rays form the floor of the abdominal furrows, and the tegumentary skeleton is supported in such a manner on the sides of the vertebral pieces of the internal skeleton as to form thereby hollow lobes, in which the intestines, or cæcal prolongations of the stomach, and a part of the genital organs extend. In the *Ophiuræ* the intestines are limited to the naked disc, and the articulated rows of the internal skeleton are everywhere surrounded by the tegumentary skeleton; so that the abdominal furrows are wanting in this group.

Professor Agassiz,[2] in the 'Catalogue raisonné des Échinides,' in treating of the affinities existing between the different orders of Echinodermata, combats the opinion expressed in the 'System der Asteriden.' " M. J. Müller affirms," observes M. Agassiz, " in his great work on the *Asteriadæ*, that the character which most clearly distinguishes these animals from other Echinoderms consists in an internal skeleton, a kind of vertebral column, on which the solid plates of the external skeleton are fixed. He affirms, even, that we observe nothing similar in the Echinidæ, of which the solid framework is altogether external. But this assertion is erroneous, and the learned anatomist of Berlin appears to me to have completely misunderstood the analogy which exists between the ambulacra of sea-urchins and the grooves on the under side of the rays of star-fishes. This analogy is nevertheless the most complete, for we here remark the same arrangement of the plates, the same openings for the passage of the tubular feet, the same relations with the ocular

[1] ' System der Asteriden,' p. 1.
[2] ' Annales des Sciences Naturelles,' 3me. série, tome vi, p. 309.

plate which is found at their summit, and with the masticatory apparatus which is found at their base. Notwithstanding the great number of the ambulacral plates, they likewise support this view of their comparative relations. As to the anal disc, it is much more extended; but this we can easily understand, if we recollect the extension which it presents, in the star-fishes, and the very narrow region circumscribed by the ocular and genital plates in the Echinidæ. The analogy of the star-fish and of the urchins is even so complete, that we may call the star-fishes urchins opened and flattened backwards, and, *vice versâ*, the urchins star-fishes contracted and inflated to form a sphere. This conformity of the urchins and the star-fishes makes me doubt the exactitude of observations which place the nervous filaments, which proceed to the eyes, at the inferior surface or external part of the ambulacra in the star-fishes, whilst they run along the internal surface of the ambulacral areas in the urchins."

In comparing the *Asteriadæ* and *Echinidæ*, as Blainville and Agassiz endeavoured to do, we soon perceive that the inter-ambulacral plates, observes Professor Müller,[1] instead of being analogous in the two orders, are quite differently arranged, and that on this circumstance in a great measure depends the difference between a sea-urchin and a star-fish. In the *Asteriadæ*, we must distinguish different kinds of inter-ambulacral plates from one another. Those which rest upon the external processes of the ambulacral plates have a certain peculiarity, as marginal plates of the ambulacra or *adambulacral* plates; they exactly agree in number with the ambulacral plates (fig. 7 *a*). To the second kind belong, in *Astrogonium* (fig. 7 *c*), the more or less well-marked *marginal inter-ambulacral plates* at the peripheral edge, which are sometimes in single, sometimes in double series. Between the ambulacral and marginal there are often *intermediate* inter-ambulacral plates (fig. 7 *b*). In *Astropecten* this area is exceedingly small, and is reduced to a few easily overlooked plates behind the angles of the mouth; in the pentagonal forms it is very large. In shape and size these plates often, as in *Astrogonium*, differ both from the adambulacral and from the marginal inter-ambulacral plates.

The marginal inter-ambulacral and the adambulacral plates extend to the end of the arms; the intermediate plates cease, for the most part, earlier. In those *Asteriadæ* whose arms are round, and whose margin is not developed, the series of plates which marks off the dorsal pore-area from the ventral surface is the equivalent of the marginal plates. In these forms, also, the number of the series of plates, from the groove of the arm to the pore-area, varies very greatly; in some there are only two series of plates, the intermediate plates disappearing, as in *Echinaster* and *Scytaster*, whilst in *Ophidiaster* there are many series of plates between the groove of the arm and the pore-area, the outermost of which, as adambulacral plates and marginal plates, extend completely to the extremity of the arm, the others, as intermediate rows of plates, are more or less, and, indeed, gradually, diminished. It is obvious that the inter-ambulacral plates of the sea-urchins

[1] 'Ueber den Bau der Echinodermen,' pp. 40, 42, 43.

and *Asteriadæ* are differently, and, in fact, so differently disposed, as to give rise to the main distinctive peculiarities of a sea-urchin and of a star-fish.

Still greater are the differences between the ambulacra of the *Asteriadæ* and *Echinidæ* in the vertical direction. The nervous cord and the ambulacral canal of the *Asteriadæ* lie, covered by the integument, over the mutually applied ambulacral plates, that is, upon the outer side of the vertebral processes of these plates ;[1] in the *Echinidæ*, however, they lie beneath the ambulacral plates on the inner surface of the shell. The vertebral processes of the ambulacral plates of the *Asteriadæ* are absent in most *Echinidæ ;* but in the *Cidaridæ* they have a perfectly analogous structure at the anterior extremity of the ambulacra, where the ambulacral plates on the inner side of the series of pores send off perpendicular processes into the cavity of the shell, between which lie the trunks of the ambulacral organs. The ampullæ are external. The clavate ends of a number of these processes unite to form a continuous colonnade, while they leave between their bases intervertebral passages, apertures for the branches given off by the ambulacral vessel to the ampullæ and the pores of the shell. There is no union of the vertebral processes of the right and left side. The analogy of the auricular processes at the anterior extremity of the corona of the sea-urchins with the vertebral processes of the *Asteriadæ*, which is remarked in the " Anatomische Studien über die Echinodermen" ('Archiv.,' 1850), is more apparent than universally true. The auricular processes are, indeed, in most sea-urchins, processes of the ambulacral plates, and the ambulacral organs pass between them ; but in *Cidaris* we meet with an exception, the inter-ambulacral plates giving off the auricular processes for the muscles of the jaws.

Besides *Cidaris, Clypeaster rosaceus,* and *altus* (or the genus *Echinanthus* altogether) possess that part of the ambulacral plates which is analogous to the vertebral processes of the *Asteriadæ*, in the internal table of their ambulacral plates. In this case all the ambulacral plates take a part in its formation, and the right and left portions are even united by a suture. This ambulacral floor lies, as in the *Asteriadæ*, beneath the trunks of the ambulacral vessels and nerves. On the other hand, the external table of the ambulacral plates lies over the trunks of the nerves and vessels, like the membranous covering of the ambulacra of the *Asteriadæ*. Herein we have sufficient evidence that, in fact, the structure of the ambulacra in the *Echinidæ* and *Asteriadæ* is widely different, and *Cidaris* and *Echinanthus* may be considered to furnish the key to the proper understanding of these deviations.[2]

---

[1] I doubt the accuracy of this statement, for my dissections of *Uraster rubens*, Lin., showed that the course of the principal nerve of the ray was along the middle of the upper part of the ambulacra arches, the position homologous to that which the nerve occupies in the *Echinidæ*, namely, beneath the ambulacral plates on the inner surface of the shell, as stated in the text by Müller.

[2] Johannes Müller, 'Ueber den Bau der Echinodermen,' 4to, plates, Berlin, 1854.

'Annals and Magazine of Natural History,' 2d series, vol. xiii, pp. 113-115. See a translation of parts of the above work by Professor Huxley, from which the above extract is taken.

" It is the *Asteridea, Ophiuridea,* and *Crinoidea,* known popularly as star-fishes," remarks Professor Huxley, " which depart least from this common plan, the ambulacral and antambulacral regions being in all these about equally developed, and the arms in most cases distinctly marked off from the body. But there are certain star-fishes which are nearly pentagonal : suppose one of these, as Müller suggests, to be elastic, so as to be capable of being distended with air into a globular form ; then the ambulacral region, with its five ambulacra, would occupy the entire apical hemisphere. There is no Echinoderm which exhibits this globular form, with equality of the ambulacral and antambulacral regions ; but if we suppose the ambulacral region to increase at the expense of the antambulacral, so that the latter eventually became reduced to a very small space around the apex, the result would be the form of *Echinus,* or of *Holothuria,* in which the ambulacral region greatly predominates, the arms disappear, and the ambulacra are, consequently, entirely calycine. One moiety of the *Cystideæ* are in the same predicament ; but other *Cystideæ,* such as *Echino-encrinus, Prunocystites, Cryptocrinus,* present a precisely opposite condition, the antambulacral region here extending into the close vicinity of the mouth, and greatly predominating over the ambulacral region. In the *Blastoidea* again, the antambulacral and ambulacral regions are more upon an equality, but the body is sub-cylindrical or prismatic in shape ; otherwise they would offer a close approximation to the hypothetical form, intermediate between an *Echinus* and a star-fish, mentioned above." [1]

Having thus reviewed the opinions advanced by different authors on the homology of the skeleton of the ASTERIADÆ as compared with the test of the ECHINIDÆ, it only remains for me to state as briefly as possible the views on this subject which I have for nearly thirty years been in the habit of teaching in my lectures on comparative anatomy. I regard the valley in the centre of the under side of the rays in star-fishes, through which the tubular retractile feet pass, as homologous to the ambulacral areas and poriferous zones in the ECHINIDÆ ; the ossicula forming the sides and upper surface of the rays of the star-fish as the homologues of the inter-ambulacral plates of the ECHINIDÆ, greatly modified for a special function. In order to show the relation of these parts to each other, I take a moderate-sized *Uraster rubens,* Lin., dead some hours, and quite flaccid, and dissect out a circle of the integument in the centre of the upper surface of the disc, including therein the madreporiform tubercle, vent, and genital pores ; the part thus removed will represent the anal area. With a pair of scissors I then lay open the upper surface of all the rays by a straight incision down the middle, from the circumference of the anal circle to the extreme point of the ray, and, folding down the two flaps thus produced from each ray into the inter-radial spaces, with a needle and thread sew the lateral flaps from the adjoining rays together ; when the whole of the flaps are thus united, I raise the border of the flat disc and form the whole into a globular shape, taking care to make the extreme points of the rays, with their eye-spots, touch the margin of the anal circle, which must be

---

[1] 'Medical Times and Gazette,' new series, No. 332, p. 463.

elevated likewise to a sufficient height, by cutting across the sand canal and other adhesions to meet the ends of the rays, and thereby form a globe. When the parts are all thus adjusted, it will be seen that they hold the same relation to each other in the ASTERIADÆ as they do in the test of the ECHINIDÆ. Thus the two flaps from each ray, when united by suture with the flaps from the adjoining rays, represent the. wide inter-ambulacral areas, with their zigzag sutures in the centre, and the surface of both being armed with spines increases the analogy. The narrow ambulacra, with their two or four rows of suckers, are undoubtedly the homologues of the ambulacral areas and poriferous zones in the *Echinidæ*. The anal area, containing the madreporiform tubercle, genital pores, and the vent, in star-fishes, represents•the apical disc formed by the ovarial plates, anal opening, and madreporiform body, in urchins. The five eye-plates at the ends of rays, in star-fishes, will fit into the margin of the circular area, when the extremity of the rays are made to approximate this part, by folding up the flat disc and converting it into a globe, which is the same singular position they occupy in the test of the *Echinidæ*. The mouth-opening will be obviously the same in position in the under side of the body in both orders.

If this demonstration is satisfactory, it is clear that we must not seek the homology between the star-fish and sea-urchin by inflating the body of disciform species, and thus making them assume globular forms, as suggested by Müller, but by placing the homologous parts in the same relation they hold to each other in these two orders of Echinodermata, always recollecting that the test of the *Echinidæ* forms a hollow globe in which the viscera are enclosed, whilst the skeleton of the *Asteriadæ* is a stelliform disc, into each ray of which a portion of the viscera is prolonged. By incising the rays down the centre of their upper surface, and folding down and uniting by suture their sides together, we produce, very clumsily it is true, the same conditions so beautifully provided in the *Echinidæ*, and reduce to a demonstration the homology of the several parts of which the body of the star-fish is composed. I have selected *Uraster rubens* for illustration, because the flexibility of the rays enables one to operate upon it easily with the scissors; but if my reasoning is correct, the observations which apply to this species will hold true with all the others, if they admitted of similar anatomical manipulation.

The Madreporiform Body.

The ASTERIADÆ, in common with the ECHINIDÆ, possess a madreporiform body; which is situated in an excentral position on the upper surface, between two of the rays (fig. 13). From the spongy plate a canal descends towards the mouth. In the *Echinidæ*, the madreporiform body always occupies the right antero-lateral ovarial plate; and as I have shown that the single ambulacral segment represents the anterior part of the body in the sea-urchins, for this reason it is inferred, that the single ray to the left of the madreporiform body in the species possessing a single plate forms the homologous part of the animal in the star-fishes.

In all the ECHINIDÆ the madreporiform body is single, and rests on the upper surface of the right antero-lateral ovarial plate; sometimes, however, it extends over the other elements of the apical disc, and surrounds them with its spongy structure. In the ASTERIADÆ the madreporiform plate is likewise for the most part single, but there are many species in which two, three, or more plates are found. It has been assumed that an increase in the number of the plates bears a certain relation to the number of the arms; observation, however, has proved that this is not the rule in all the many-rayed forms. *Solaster* and *Luidia*, for example, which possess numerous rays, have the madreporiform plate single, whilst in some five-rayed *Ophidiasters* the plate is double. With an increased number of arms there is, in some genera, a corresponding increase in the number of the madreporiform plates. Thus, in *Asterias Helianthus*, Lamk., which has from thirty to thirty-six rays, the madreporiform plate consists of many pieces; and *Uraster tenuispina*, Lamk., which has from six to eight rays, possesses two or three plates. The genera *Ophidiaster* and *Echinaster*, in general, have more than one plate. In *Ophidiaster multiforis*, Lamk., the individuals with five rays have two plates, whilst those with six rays have three; and in *O. diplax*, *O. ornithopus*, *O. Ehrenbergii*, all five-rayed species, there are two plates in each. The remarkable *Echinaster solaris*, so beautifully figured by Ellis,[1] has twenty short rays armed with very long spines, and around the circumference of the anal area sixteen hemispherical, madreporiform plates are figured; in other individuals of the same species, examined by Müller and Troschel,[2] the number of plates was not so great; one specimen with fourteen rays had five, and another with sixteen rays had six plates; and *Echinaster Eridanella*, Valenc., with six rays, has two madreporiform plates. In those genera, therefore, the number of the plates appears to augment with the number of the rays.

The Tegumentary Appendages.

The tegumentary membrane in the ASTERIADÆ is provided with different kinds of appendages, as *spines, granules, paxillæ,* and *Pedicellariæ,* each of which requires a separate notice.

The *spines* are calcareous pieces of various forms and sizes; they are in general attached by their base, and often destitute of the kind of articulation seen in the spines of ECHINIDÆ. In *Uraster*, Pl. I, fig. 2, they are sharp, prickly processes arranged in rows, with more or less regularity, along the upper surface, sides, and base of the rays. In *Astropecten* (fig. 3) and *Luidia* (fig. 6), they are in the form of long, tooth-like spines which project from the sides of the marginal plates. In *Pteraster* (fig. 5) they fringe the borders of the rays, form fan-like semicircles near the ambulacra, and arm the upper

---

[1] 'Natural History of Zoophytes,' p. 206, Pl. 60, 61, 62.
[2] 'System der Asteriden,' p. 25.

surface of the lobes with thorny prickles. In *Echinaster* they are developed into long defences, and thickly set together on all the surface of the body. In *Oreaster* (fig. 13) they are thick calcareous pieces, which rise in various forms from the surface of the ossicles. In *Astrogonium equestris* the smooth spines project from the centre of a nearly circular plate, around the border of which is a circle of granules; the intermediate spaces are filled with tubercles, among which valve-shaped Pedicellariæ are scattered (fig. 8). Besides the spines disposed on the sides and upper surface of the rays, there are others which, in general, have a very regular arrangement, and form consecutive rows on each side of the ambulacral valleys.

The *granules* are fine, calcareous, wart-like processes, which grow from the surface of the integument, and cover all the rays in *Ophidiaster* and *Scytaster*. In other genera they are much more limited in their distribution, and occupy the inter-spinous spaces on the surface of the rays.

The *paxillæ* are formed of processes of the integument, which rise like short stems in regular order from the surface of the ossicules ; each stem carries a crown of short, bristly spines, as in *Solaster* (fig. 4). In *Scytaster* they are distributed over the discal membrane, and arranged in lines on the sides and upper surface of the rays. In *Astropecten* (fig. 3), *Luidia* (fig. 6), *Ctenodiscus*, and *Archaster*, they fill the entire space on the upper surface within the area circumscribed by the marginal plates.

The *Pedicellariæ* are small, pincers-like bodies, supported on slender, flexible stems, and found in considerable numbers around the bases of the spines and on the membrane surrounding the mouth. They were first observed by Müller[1] on the test of a sea-urchin (*Echinus sphæra*), and described by him as *Epizoa*. Lamarck[2] classed them with the Polypes, and Cuvier[3] doubtfully adopted the same view, as also Schweigger;[4] whilst Munro, Oken, Delle Chiaje, Sharpey, Valentine, Sars, Müller and Troschel, and Forbes considered them as tegumentary appendages of the animals on which they are found.

In *Uraster rubens*, Lin., groups of these pedicellated, pincers-like bodies are seen clustering around the base of the spines, each consisting of a membranous stem, surmounted by a pair of calcareous forceps not unlike the miniature claw of a Crustacean ; when alive and active, if a fine needle is introduced between their expanded blades, they close upon the foreign body, and grasp it with force. Professor Forbes[5] examined the *Pedicellariæ* in this star-fish, and observed that "those on the body and upper spines differ in shape from those on the spines which are arranged on the sides of the ambulacral valleys. The former are much shorter and blunter in their blades than the latter. The calcareous forceps of which their heads consist are imbedded in an integument of a soft, granular

[1] 'Zoologia Danica.'

[2] 'Animaux sans Vertèbres,' 1st ed., vol. ii, p. 63.

[3] 'Règne Animal,' 2d ed., vol. iii, p. 297.

[4] 'Handbuch der Naturgeschichte.'

[5] 'History of British Star-fishes,' p. 98.

tissue, which envelops the forceps when closed; and this apparatus is mounted on a bulging body of a similar substance, which crowns the round, flexible, and contractile peduncle, sometimes simple, sometimes branched, each branch having a similar termination. I could detect no evidence of vibratile cilia on their stalks; but there appeared to be ciliary motions within the blades. When the star-fish is alive, the *Pedicellariæ* are continually in motion, opening and shutting their blades with great activity; but when cut off, they seem to lose their power." The *Pedicellariæ* observed on certain Echinodermata have been most carefully examined by Sars,[1] and I shall enrich this branch of the subject with that accomplished naturalist's observations on these remarkable appendages of the tegumentary membrane.

"In examining *Echinus sphæra*," says Sars, "I found upon it all the three sorts of *Pedicellaria* described by Müller, viz., *P. tridens*, *P. triphylla*, and *P. globifera*. Besides what Müller states in regard to *P. tridens*, I will make the following remarks:—Internally, there is a hard stem, which is enclosed by a strong, transparent skin, like a sheath. It is thickest at the upper and lower ends, and reaches from the neck, as it is called, to the base, where it, remarkably enough, is fixed and jointed to an exceedingly small barb projecting from the sea-urchin's shell. This circumstance, which is invariable in the *Pedicellaria*, seems not to have been sufficiently attended to. The three teeth are concave on the side, turned inwards, angular, and furnished with small teeth on their edges. They are hard and calcareous; when viewed through a microscope, they are seen connected with very small globules arranged in rows. The stem is also calcareous, yet it can be slightly bent without breaking. The neck is nearly as thick again as the stem; it is fleshy, transparent, and very flexible.

"The motions observed in the *Pedicellariæ*, when irritated, are that the teeth close and squeeze pretty firmly; in this way, by inserting the point of a pin between them, after the *Pedicellaria* was torn off, I could draw it out of the water; further, that the neck bends and inclines to all sides, and can even contract a little, in doing which transverse wrinkles are formed on it; and, lastly, that the stem, itself inflexible, may bend along with the whole *Pedicellaria* to the side. The form called *Pedicellaria globifera* by Müller has a head consisting of three outspread flaps, standing nearly horizontally. Each of these flaps is oval, very convex externally, and concave internally, and at the upper end slightly indented, and provided with a sharp point, somewhat bent. From the indentation runs a raised stripe or rib longitudinally downwards through the flap. On the inner side of these flaps, at their base, is seen an oval and apparently calcareous leaf.

"The stem, which is similarly constituted with that of *P. tridens*, proceeds directly from the head (there is no neck in this species), is small above, and thicker below, until at the bottom it completely fills the hollow of the sheath which encompasses it.

---

[1] 'Ueber die Entwickelung der Seesterne.'—Müll. Arch. 1842, p. 330. 'Ueber die Entwickelung der Seesterne.'—Wiegm. Arch., 1844, ii, p. 169, fig.

" With regard to the motions of these *Pedicellariæ*, they not only quickly open and shut the three flaps, but can also turn the head to the different sides, and up and down, and that very quickly.

" Müller says, regarding *P. tridens:*—' Variat absque aristis, an perditis?'   Of such I have also found a large number of specimens ; but I scarcely believe that they belong to *P. tridens*, since the teeth of these last are fixed so firmly that they could scarcely fall off.   Either they are a separate species or a variety of *P. triphylla*, which they resemble in every point, except that the three flaps are broad at the bottom and small at the ends.   These flaps seemed there also to be calcareous, and consisted of many small globules, which were arranged in transverse rows, clearly separated from each other by a light transparent line.   Such a line also ran longitudinally down the flap.   In *P. triphylla* the flaps are not obtuse, but a little rounded, and have, like the foregoing, globules extending in rows.   If we now consider the construction of the *Pedicellariæ* and their manner of life as a whole, we can scarcely believe them to be anything but organs of the sea-urchin."

The following reasons seem to prove the accuracy of this opinion :

" 1st. In all sea-urchins, without exception, are found *Pedicellariæ*, and under the same circumstances; which would certainly not always be the case if they were parasitical animals,—just as *Lernææ* are not always found in all fishes, &c.

" 2d. The hard calcareous teeth or plates, and the internal stem, also calcareous, and often filling up alone the sheath, which are found in all *Pedicellariæ*, bear a greater resemblance to an Echinus spine than to any animal of the Polype kind.   There is neither opening, nor mouth, filaments, &c.

" 3d. The *Pedicellariæ* are firmly fixed in the skin which envelops the whole sea-urchin, upon a very small projecting knob of the shell, to which knob they are very strongly attached, but yet moveable, like the prickles of the sea-urchin ; the under surface of the stem of a *Pedicellaria* being somewhat hollowed and articulated with the knob. When a *Pedicellaria* is torn out, it is observed that the sheath or skin connecting the stem is torn at the lower end, which, doubtless, is a consequence of its connection with the skin, with which the shell of the sea-urchin is covered, and which, when the *Pedicellaria* is torn out, must be rent.

" 4th. When the skin of the sea-urchin or a single *Pedicellaria* is irritated— for example, with a pin—the surrounding *Pedicellariæ*, which stand in a wide circle, invariably bend themselves quickly towards the irritated part.   This phenomenon, which I have often observed, shows clearly an organic connection between the *Pedicellaria* and the skin of the shell of the sea-urchin.

" The same thing precisely is observed with the spines."

On the use of these bodies, M. Sars continues : " Perhaps Nature, who has so abundantly provided the sea-urchin with such an astonishing number of feet and prickles, has also given the *Pedicellariæ*, as a sort of antennæ to seize the small animals which serve for its sustenance, partly to lay hold of whatever might approach their sensitive skin which covers.

the surface of the shell, and thus, in conjunction with the prickles, protect it from injury."
Professor Edward Forbes, in discussing the function of the *Pedicellariæ*, says, " If they
be not distinct animals, as Müller fancied, for what purpose can they serve in the economy
of the star-fish? If they be parasites, to what class and order do they belong ?—what
is their nature, and what is their food ? Truly, these are puzzling questions. These
organs or creatures have now been known for many years, have been examined and ad-
mired by many naturalists and anatomists, have been carefully studied and accurately
delineated, and yet we know not what they are. This is but one of the mysteries of
natural history—one of those unaccountable things which we know and know not—of
those many facts in nature which teach us how little is man's knowledge, and how won-
drous and unsearchable is God's wisdom. It is folly and vanity to attempt to account
for all facts in nature, or to pretend to say why the great Creator made this thing, and
why He made that, and to discover in every creature a reason for its peculiar organization.
It is but another form of the same vanity, having satisfied itself of the discoveries it has
made, to pretend to praise the all-wise Maker's wisdom in so organizing His creatures.
That God is all-wise is a revealed truth ; and whether the organization before us seem
excellent or imperfect, it matters not—we *know* it is perfect and good, being the work of
an all-wise God."[1]

The Vent.

It was long believed that the *Asteriadæ* were destitute of an anal opening, but a more
careful study of the organization of these animals has proved this was an error. Baster,[2]
in reference to *Uraster rubens,* wrote : — " Utrumque genus (Echinorum et Stellarum
marinarum) os inferne, et ad excrementa ejicienda aperturam superne habent."

Janus Plancus[3] observed :—" Præterea hæ stellæ anum in medio oppositum ori, uti
Echini, veluti umbilicum quemdam gerunt et in acumen attollunt." Müller[4] describes, in
the *Asterias militaris,* a central spot as " macula verruciformis," and says as this spot is
not perforated, therefore Baster's description of the anus could not be correct.

Tiedemann, in his great work, denied Baster's statement, and treated his observation
as a mistake ; since the publication of that treatise, the same opinion has been expressed
in nearly all modern zoological works. This discrepancy about an anatomical fact has
arisen from the error of deducing general conclusions from limited observations ; for it
now appears that, of the eighteen genera of star-fishes described by Müller and Troschel

---

[1] Forbes, ' British Star-fishes,' p. 98, 99.
[2] Baster 'Opuscula subseciva,' p. 116.
[3] ' Epistola de incessu marinorum Echinorum. Opuscula Instituti Bononiensis,' tom. v, pars i, p. 245.
[4] ' Zoologia Danica,' cxxxi, p. 14.

fifteen of these possess a vent, and only three are destitute of that aperture. The star-fish which Tiedemann dissected and illustrated with magnificent plates (*Astropecten aurantiacus*, Lin.) belongs to the latter group; whilst Baster's observations were made on *Uraster rubens*, Lin., which is classed with the former, and possesses an anal opening. The vent is sub-central, and lies in general at the left side of the madreporiform plate, sometimes surrounded, as in *Oreaster reticulatus*, by a circle of small wart-like tubercles. The annexed fig. 14 shows the size and position of the vent aperture in this large star-fish.

FIG. 14.

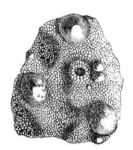

A portion of the disc, with the vent in *Oreaster reticulatus*.

STRATIGRAPHICAL DISTRIBUTION OF THE FOSSIL ASTERIADÆ.

It was long supposed that the *Crinoideæ* were the only representatives of the Echinodermata in the Palæozoic rocks, but recent researches in the Silurian strata of England, Wales, Ireland, and North America, have led to the discovery of *Asteriadæ* in several stages of these ancient formations. Professor Sedgwick and Mr. Salter, in 1845, found *Palæaster obtusus*, Forb., in the ash-bed west of Bala Lake, and the same species was afterwards discovered in 1846 by Sir Henry de la Beche, Captain James, R.E., and Professor Forbes, in the fossiliferous slates of Drumcannon, near Waterford, which are of Bala and Caradoc age, as shown by the fossils of all their fossiliferous portions. The Lower Silurian rocks of North America have likewise yielded *Asteriadæ*. Mr. James Hall has described and figured *Palæaster Niagarensis*, Hall, from the Trenton limestone of the State of New York, and Dr. Billings has described many new forms from the Lower Silurian rocks of Canada, and figured the same in the third decade of Canadian organic remains of the 'Geological Survey of Canada.'

The Upper Silurian strata of Westmoreland and Shropshire have likewise lately been found to contain some beautiful little sea-stars in fine preservation, and Mr. James Hall has discovered several new forms in the Upper Silurians of the United States.

The Palæozoic *Asteriadæ* have been studied with great care by the late Professor Forbes, Mr. James Hall, Mr. Salter, and Dr. Billings. All the species belong to extinct genera, which present many singular modifications of structure, and afford interesting points of comparison with some living forms. The following genera have been proposed for the reception of these Palæozoic *Asteriadæ*; the diagnosis of each is given in the words of their respective authors.

Genus 1. PALÆASTER, *Hall.*—Arms thick, convex, short, or moderately elongate, and formed of many rows of small, spinous ossicles on the upper surface; ambulacra deep, with transverse ossicula, and a single row of adambulacral plates. No disc plates between the rays; madreporiform tubercle small and single. This genus ranges from the Lower Silurian to the Carboniferous strata.

Genus 2. PALASTERINA, *M'Coy.*—Pentagonal depressed; the arms a little produced, with three or five principal rows of tubercles above, combined with a plated disc which fills up the angles; ambulacra rather shallow, formed of subquadrate or slightly transverse ossicles, bordered by a single row of large, square-shaped plates, the lowest of which are large and triangular, and support combs of spines. Upper Silurian rocks.

Genus 3. STENASTER, *Billings.*—No disc; rays linear, lanceolate, or petaloid; ambulacral grooves bordered by solid, oblong, or square adambulacral plates; five pairs of triangular oral plates; two rows of ambulacral pores. Upper surface of the disc and rays covered with small plates, which appear to be tubercular, and not closely fitted together. The generic name is from *stenos*, narrow, in allusion to the contracted body. Lower Silurian.

Genus 4. PETRASTER, *Billings.*—This genus has both marginal and adambulacral plates, with a few disc-plates on the ventral side. The general form is deeply stellate, and the rays are long and uniformly tapering. Generic name from *petra*, a stone. It differs, according to Dr. Billings, from *Palasterina* by the presence of large marginal plates outside of the disc-plates, and still more from *Stenaster,* which has neither discal nor marginal plates. It is, however, allied to *Astropecten.* Lower Silurian.

Genus 5. PALÆOCOMA, *Salter.*—Flat; all the centre of the disc above membranous, with scattered, star-like calcareous spiculæ; the angles filled up with a similar membrane. Arms formed of several rows of quadrate, reticular ossicles, the external rows fringed with spines. Beneath, the ambulacra are narrow and very shallow, the ossicles square or even elongate, and placed alternately. Two rows of bordering plates, the inner row square and without spines, the outer row oblique and fringed with combs of very long spines; a loosely reticulated membranous web between the arms. Upper Silurian.

Genus 6. PROTASTER, *Forbes.*—Arms elongate, extending much beyond a circular, closely reticulated disc. Arms formed above of two rows of plates, deeply sculptured and spinous at the edge; beneath, of two rows of elongate ambulacral ossicles, bordered by a row of large spinous plates. The basal ossicles of the ambulacra, bordering plates, and disc, combined to form a petaloid mouth below. Upper and Lower Silurian rocks.

Genus 7. PALÆODISCUS, *Salter.*—Arms not produced beyond the large, *plated* pentagonal disc, nor distinguishable from it above; ambulacra formed beneath of crowded, transverse ossicles, the basal joints of which are greatly enlarged, thickened, and placed in vertical pairs to form the mouth.

Genus 8. TÆNIASTER, *Billings.*—Body deeply stellate; no disc or marginal plates; rays long, slender, flexible, and covered with small spines; two rows of large ambulacral pores; adambulacral plates elongated and sloping outwards so that they partly overlap each other; adambulacral ossicula contracted in the middle, dilated at each end. Generic name from *tænia*, a riband.

PALÆASTER.—*Hall.*

Palæaster asperrimus, Salter (Annals and Mag. of Nat. Hist., 2d series, vol. 20, p. 325, pl. ix, fig. 1) fig. 15[1].

FIG. 15.

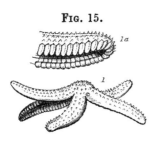

Rays five, short, round and obtuse; upper surface convex (1), and ornamented with many longitudinal rows of prominent tubercles; a single madreporiform body at the angle between two rays. Ambulacra wide; grooves deep, bordered by two rows of large, transverse, marginal, adambulacral ossicula, with acute ridges on their under side (1 *a*).

Locality.—Collected by the Geological Survey in the Caradoc or Bala sandstones, near Welchpool, N. Wales.

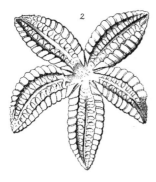

Fig. 1. *Palæaster asperrimus*, Salt.
Fig. 2. *Stenaster Salteri*, Billings.

Palæaster obtusus, Forbes. Mem. Geol. Surv., Decade 1, pl. i, fig. 3, 1849.

"Body rather broad, convex above, spinosely reticulated; spines very short, and probably grouped in tufts. The arms are short, convex above, broad, oblong, and obtuse. Their under surfaces exhibit oblong, rather broad, ambulacral plates, gradually decreasing in size towards the tips of the arms, but nearly equal for about two thirds of their length; the ambulacral sulcus between them is rather broad. The largest specimen examined measured an inch and a half across."

Locality.—First found in Lower Silurian rocks at Drumcannon, near Waterford, in 1846, by Sir Henry de la Beche, Captain James, R.E., and Professor Forbes, and by the Geological Surveyors in the ash-bed of the Bala rocks, West of Bala Lake, North Wales. In the Irish locality it was associated with *Phacops Jamesii* and numerous *Orthides*. In the Welsh, with Trilobites of the genera *Asaphus*, and *Homalonotus*. Brachiopoda of the genus *Orthides*, and numerous stems of *Encrinites*.

Palæaster coronella, Salter.   Ann. and Mag. Nat. Hist., 2d series, vol. xx, p. 326.

" A small species, with four rows of tubercles on each arm, and a close corona of six tubercles."

Locality.—Gunwick Mill, Malvern (in the May Hill Sandstone).

Palæaster Ruthveni, Forbes.   Mem. Geol. Surv., decade i, pl. i, fig. 1.

" Body very small, in the centre of five tapering linear lanceolate, rounded rays, subcarinated on their upper surfaces, five times as long as the disc is broad.   The upper surfaces of both rays and disc are reticulated, indicating a structure which originally, in all probability, consisted of spines grouped in tufts.   The under surfaces are marked by the impressions of a double series of ambulacral articulations, each slightly curved.   Both these structures are present in some existing antarctic forms of *Uraster*.   The largest specimen examined measured three inches and a half across, from arm-tip to arm-tip."

Locality.—At Scalthwaiterigg, and Highthorns, Westmoreland; in Ludlow Rocks. The original specimen is in Professor Sedgwick's collection.

Palæaster hirudo, Forbes.   Mem. Geol. Surv., decade i, pl. i, fig. 4.

"Body very minute, about a fourth as broad as the rays are long; rays tapering and linear-lanceolate, contracted at their bases, pointed at their extremities.   Their upper surface clothed with bundles of spines arranged in regular rows, and so placed that each ray seems to be marked by three or four longitudinal furrows, crossed at regular intervals by transverse grooves.   Under surface with short ambulacral plates and broad avenues.   The largest specimens do not measure more than an inch across."

Locality.—Gregarious in Ludlow Rocks, at Pottersfell, near Kendal, Westmoreland.

Palæaster Niagarensis, Hall.   Palæontology of New York.

Body stellate, arms tapering, ambulacra wide, under surface of the rays with large marginal adambulacral plates, and five oral plates.

Locality.—Trenton Limestone, Lower Silurian series, New York.

A small *Palæaster* has been found at Braunton, near Barnstaple, N. Devon, in the

4

lowest beds of the carboniferous rocks ; these beds, Mr. Salter kindly informs me, are called by Mr. Jukes and himself the " Coomhola grits" in Ireland, and which Sedgwick and Murchison called the " Marwood beds" in N. Devon : they are neither Devonian nor Carboniferous, but lie on the confines of both. This asteroid is not yet described.

PALASTERINA, *McCoy.*

Palasterina primæva, Forbes. Mem. Geol. Surv., decade i, pl. i, fig. 2.
— — Salter. Ann. and Mag. of Nat. Hist., 2d series, pl. ix, fig. 2, from which fig. 16 *a* is copied.

Body broad, pentagonal, produced at the angles into five short, lanceolate, or elongato-triangular, pointed arms, which are each about two thirds as long as the breadth of the disc. Surface of the disc convex above (*a*), as well as the arms tuberculated and reticulated, exhibiting traces of having been covered by tufts of short, blunt spines. Madreporiform plate and vent, both subcentral. Beneath, nearly flat ; the inter-ambulacral spaces reticulated like the upper surface ; the ambulacra composed of broad, oblong, geniculated plates (*b*), of which there are about twenty in a row. The largest specimens examined had attained the dimensions of an inch and a half in diameter, from arm-tip to arm-tip. This star-fish has many affinities with *Asterina* or *Asteriscus*. Mr. Salter has discovered that the " basal or angle ossicula are enlarged, three-cornered, and furnished with a pyramid of spines, pointed inwards. The upper surface is roughly tuberculate, and possesses short tufts of spines."

FIG. 16.

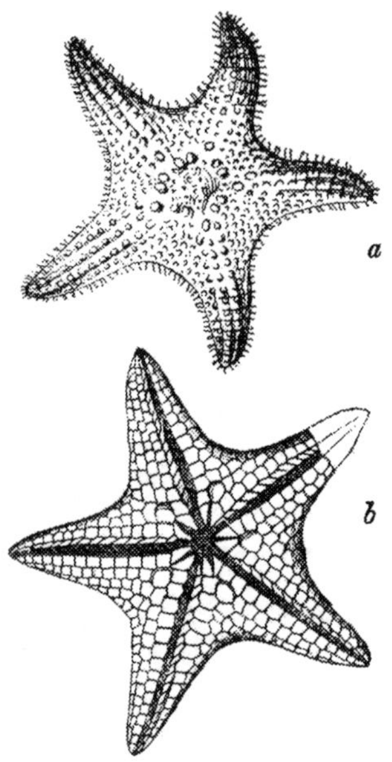

a

b

Palasterina primæva, Forbes, fig. *a.*
„ *stellata*, Billings, fig. *b.*

Locality. — Underbarrow, near Kendal, Westmoreland. It is found in a thin, subcalcareous band of Ludlow Rock, loaded with *Trilobites* and *Encrinites*.

Palasterina antiqua, Hisinger. Leth. Suec., tab. xxvi, fig. 6, p. 89.

Locality.—Mount Hoburg, Gothland, Sweden ; in Ludlow Rocks.

The Canadian species of *Palasterina* I have derived from Dr. Billings' valuable paper on the *Asteriadæ* of the Lower Silurian Rocks of Canada, in the third decade of Canadian Organic Remains.

Palasterina stellata, Billings. Geol. Surv. of Canada, Organic Remains, decade iii, pl. ix, fig. 1, p. 76. Fig. 16 *b* copied from this plate.

Description.—Pentagonal; disc about one half of the whole diameter; ambulacral grooves narrow and deep, bordered on each side by a row of small, nearly square adambulacral plates; a second row, consisting of disc-plates, extends nearly to the end of each ray, the remainder of the disc covered with smaller plates. All these plates are solid and closely fitted together; the disc-plates in the angles in contact with the oral plates are much larger than any of the others (fig. 16, *a*).

In the only specimen in the collection, the length of the rays, measured along the ambulacral grooves, is three lines; number of adambulacral plates on each side of the grooves, sixteen; the rays diminish somewhat rapidly in size, and terminate in a rounded point; diameter of the disc four lines. The plates are all a little worn, so that the character of their surfaces cannot be observed; they were probably nearly smooth.

Locality and Formation.—City of Ottawa; Trenton Limestone. Collected by Dr. Billings.

Palasterina rugosa, Billings. Geol. Surv. of Canada, Organic Remains, pl. ix, figs. 2 *a, b, c.,* p. 77.

Description.—Two inches in diameter; rays five, acute at their apices, and rapidly enlarging to a breadth of four lines at the disc, which is eight lines in width. The specimen shows the upper side of the fossil only. Some of the plates are absent from the centre of the disc, but those which remain are very prominent in their centres, and roughly ornamented with four or five deep crenulations or furrows from near the centre to the edges, producing a star-like appearance, resembling a half-worn plate of *Glyptocrinus decadactylus;* their diameter is from one to two lines.

The rays are composed (at least, the backs and sides of them) of four rows of plates, which are so very prominent that they appear to be almost globular, and even pointed in their centres; the central rows are the smallest; the first four plates of the outer row occupy three lines in length, and of the inner rows nearly as many. Towards the point of the arm all diminish rapidly in size. Beneath the outer rows two others can be seen, which are probably the outer marginal plates of the under side, corresponding to those of *P. rigidus.*

Locality and Formation.—Charleton Point, Anticosti; Hudson River group. Collected by J. Richardson.

[1] 'Geological Survey of Canada.' Canadian Organic Remains, decade iii, p. 76, pl. ix, fig. 1-15.

[2] Ibid, p. 77, pl. ix, fig. 2.

STENASTER.—*Billings.*

Stenaster Salteri, Billings.   Geol. Surv. Canada, Organic Remains, pl. x, fig. 1 *a, b,*
    p. 78.

Description.—This species has rather short broad rays, which are narrower where they
are attached to the very contracted body, than they are at about the centre of their length.
In consequence of this form, the sides of the rays are not parallel, but a little curved
outwards.   As, however, only two specimens have been collected, and both appear to be
a little flattened by vertical pressure, it may be that this leaf-like shape of the rays is
accidental, and that in perfect specimens they taper uniformly from the body outwards.
The adambulacral plates are oblong, and the sutures between them are nearly at right
angles to the ambulacral grooves; those next the body are a little sloping outwards.
Their length is about twice their breadth, and they are so disposed that the greater
dimension is transverse, or at right angles to the groove; the extremities which lie next
to the grooves are angular, and some of them appear to have the contiguous pores partly
excavated in them.   The oral plates are acutely triangular, the sharpest angle being
towards the mouth.   The plates are smooth.   The ambulacral pores are very large, and
the ossicles are much contracted in the middle, and greatly expanded along the median
line of the bottom of the groove.

    The most perfect specimen is one inch in diameter, measured between the tips of
the rays; diameter of disc, three lines; width of ray at mid-length, two lines and a half
Dedicated to J. W. Salter, Esq., Palæontologist of the Geological Survey of the United
Kingdom.

    Locality and Formation.—Belleville, Canada West; Trenton Limestone.   Collected
by Dr. Billings.

Stenaster pulchellus, Billings.   Geol. Surv. of Canada, Organic Remains, decade iii,
    pl. x, fig. 3, p. 79.

Description.—Rays long, slender and sub-cylindrical; adambulacral plates, transversely
oblong; grooves narrow; dorsal plates small and tubular.   Diameter of the only specimen
in the collection two inches and one fourth, measured between the tips of the rays; arms
one inch in length, and two lines and a half in width at the base; disc three lines and a
half in diameter.

    Locality and Formation.—Ottawa; Trenton Limestone.   Collected by Dr. Billings.

PETRASTER.—*Billings.*

Petraster rigidus, Billings.   Geol. Surv. of Canada, Organic Remains, decade iii, pl. x, fig. 3, p. 80.

Description.—This species has much the aspect of an *Astropecten ;* the disc is one fourth the whole diameter, the rays rather slender, and uniformly tapering ; the angles between the bases of the rays rounded.   The plates, which appear to be adambulacral, are quadrate and a little convex ; the marginal plates oblong, and also convex ; the disc-plates consist of three at each angle, and a single row on each side of the ray, but extending only one third or one half of the length of the ray ; they all lie between the marginal and adambulacral plates.   The specimen figured was about two inches in diameter when perfect ; width of disc, half an inch ; and of rays at the base, about three lines.

Locality and Formation.—Trenton Limestone, Ottawa. Collected by Mr. J. Richardson.

PALÆOCOMA,[1] *Salter.*

The genus *Palæocoma* was proposed by Mr. Salter[2] to receive an interesting group of star-fishes, lately discovered at Leintwardine, Shropshire, in the thin flagstones of the Lower Ludlow Rock, which were there associated with *Pterygotus, Ceratiocaris* of several species, and *Limuloides,* a genus apparently allied to *Limulus,* together with new *Crinoidea* and *Polyzoa,* and many of the more common *Brachiopoda* and *Graptolites* of the Lower Ludlow series, which are here overlain by layers of Aymestry Limestone, full of *Pentamerus Knightii.*

Palæocoma is characterised by the elongated form of the narrow ambulacral ossicles, which are bordered by a double row of marginal adambulacral plates, the outer row supporting combs of long spines ; " the spines are often so long as to form a complete fringe, and in one species the disc is equally spiniferous.   In the curious sub-genus, *Bdellacoma,* they are short, and intermixed with some larger clavate spines on the upper surfaces ; and in *Rhopalocoma,* which may hereafter have to be separated as a distinct genus, the hair-like spines are all absent, and clavate ones take their place."[3]   *Palæocoma* is nearly related to *Pteraster militaris,* now living in the seas of Greenland and Spitzbergen, in the spinigerous character of the adambulacral plates, but differing from it in the manner

---

[1] The name *Palæocoma* was, unfortunately, previously given by the late M. A. d'Orbigny to a genus of the *Ophiuridæ.*

[2] 'Annals and Magazine of Natural History,' 2d series, vol. xx, p. 327, pl. ix, fig. 3.

[3] Ibid., p. 327.

the spines are arranged thereon; *Pteraster* having combs of spines united by a membrane on the inner row of plates, and long, stiff, fin-like spines on the outer row, whilst *Palæocoma* has the outer row spinigerous, and the inner row destitute of these appendages. The three species of *Palæocoma* described by Mr. Salter, and the two subgenera, *Bdellacoma* and *Rhopalocoma*, were all obtained from the Lower Ludlow Rocks at Leintwardine.

Palæocoma Marstoni, Salter.   Ann. and Mag. Nat. Hist., vol. xx, p. 328, pl. ix, fig. 3.

Arms lanceolate, obtuse at the apex, united throughout by a delicate reticulate membrane clothed with spines; mouth opening wide, angle ossicles double; ambulacral grooves narrow, and bounded by two rows of small, quadrate ossicles; the margins of the rays are formed by two series of adambulacral plates, armed with regular oblique combs of long rigid spines, which form a conspicuous spinous fringe on the margins of the arms. Mr. Salter says, " that shorter spines, set at a wide angle from the margins of the arms, distinguish this from the next species. The beautiful and delicate web-like disc between the arms bears short spines also; it is sometimes expanded, as in fig. 17, but more generally contracted, so as only to make the arms a little broader, and give it a blunt appearance. The mouth is wide, of a true pentagonal shape, and with rather strong bordering plates, of which the triangular adambulacral plates are most conspicuous. The calcareous stellate spiculæ which dot over the thin disc are easily seen through the opening of the mouth."

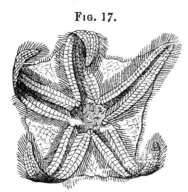

Fig. 17.

Palæocoma Marstoni, Salter.

Locality.—Church Hill, Leintwardine, Shropshire, in Lower Ludlow Rock.

Palæocoma Colvini, Salt.   Resembles the preceding species, but is remarkable for the length of its hair-like spines, which exceed those in any known star-fish, recent or fossil; it is found with the preceding.

Palæocoma Cygnipes, Salt.   Arms elongate, marginal spines short, inter-brachial membrane delicately expanded, mouth opening small; found associated with the preceding.

Mr. Salter forms sub-genera of two other species belonging to this group, which present characters differing from other *Palæocomæ*.

Sub-genus—BDELLACOMA, *Salter.*

Bdellacoma vermiformis, Salt. Arms long, bordered by short spines; ambulacral avenues wide and flat, with large alternating apertures for the suckers. It is doubtful if the avenues are bordered by more than a single row of plates; but as there is a double set of tufts of spines, this is probable. The author observes, " the main character of the species, however, and that which distinguishes the sub-genus, is the possession of scattered clavate tubercles over the upper surface. These are nearly as long as the spines. The ambulacral avenues too, appear to differ materially from those of *Palæocoma*, in which they are remarkably narrow, and the plates close, while in *Bdellacoma* they are broad, and the ossicles remote."

Locality.—Leintwardine, in Lower Ludlow Rock.

Sub-genus—RHOPALOCOMA, *Salter.*

Rhopalocoma pyrotechnica, Salt. The sub-generic and specific characters of this fine species, the author states, must be taken together, and reside in the distribution of short, broad, clavate, and compressed spines over the upper surface and margin, rather more than their own breadth apart, and set on at the intersection of the reticular meshes, which cover the arms, and the angles between the arms, but which are quite absent from the central portion of the disc. This central portion above, which corresponds to the wide aperture of the mouth on the under surface, is covered only by scattered, stellate, calcareous spiculæ of large size. A closer reticulation is found on the portions between the arms, and the meshes become square in a double row, down the middle of the rays, and appear to correspond nearly in position to the ambulacral bones of the under surface. The latter are very slender and remote, even more so than in *B. vermiformis*, and form a broad ambulacrum, with only a few reticular plates bordering it, which bear clavate spines at intervals. The mouth angles project a good deal inwards, and are armed with short combs of spines."

Locality.—Leintwardine, in Lower Ludlow Rock.

PROTASTER, *Forbes.*

The genus *Protaster* was instituted by Professor Forbes, to comprehend those Silurian star-fishes which have a small circular disc covered with squamiform plates. The arms are long, slender, and simple, and form a pentagonal rosette in the centre of the disc; they are constructed of two rows of alternating quadrate ossicles, which have their outer margin indented superiorly, to form spinigerous crests. The spines are apparently short, and not equal in length to that of an ossicle, obtuse, and few in a row. The madreporiform plate (fig. 18, *a*) has been found between two of the rays. This remarkable genus reminds us, at first sight, of the *Ophiuridæ*, to which it was referred by its author. The discovery of the madreporiform plate, however, removes it to its place among the *Asteriadæ*. The alternating quadrate ossicula of the arms connect it by osteological affinities with the *Euryalidæ*, of which it appears to have been the ancient type.

Protaster Miltoni, Salter. Ann. and Mag. of Nat. Hist., 2d. series, vol. xx, p. 330, pl. ix, fig. 4.

"Disc round, one inch in diameter, and covered with small ridged plates; the arms are wide, composed of large ossicles, which become smaller at their base of insertion in the disc above, having a wedge-shaped space between, which, joined with those of the other arms, forms a conspicuous pentagon above, equal in size to that formed by the divergent ossicles of the mouth below.

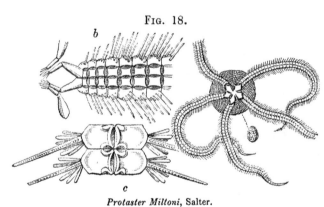

Fig. 18.

Protaster Miltoni, Salter.

b. The upper surface of a ray.
c. The under surface of a ray with the oral combs.

The arms themselves are made up of a double row of about forty pairs of squarish concave plates above, placed exactly opposite, not alternating as in other species (fig. 18 *b*). The sutures between these are deep, and the inner angles marked with a deep pit or pore, bounded by tubercles, set cross-fashion. The outer margin bears a tuft of spines, long and short. On the under surface (fig. 18, *c*) the marginal plates are highly convex, and between them lies a double row of central plates of an hour-glass shape, on the outer sides of which, and between them and the marginal row, is a large, round aperture on either side—the passage for the tubular feet. The marginal plates bear a row of spines as long as the width of the arm, and striated across (fig. 18, *c*).

The oral pentagon is made up of twenty bones, five pairs of which are the central row enlarged (fig. 18, *b*), and these diverge at a wide angle, and nearly join the neighbouring pairs; the other five belong to the lateral rows, and are linear, set parallel, and bear the conspicuous, triturating combs of spines. The oral ossicles in this species form an angular pentagon, being made up of straight pieces; in some others they are arched, and give an ogive form.

Locality.—Leintwardine, abundant of all sizes in Lower Ludlow Rock.

Protaster leptosoma, Salter. 'Ann. and Mag. of Nat. Hist.,' 2d series, vol. xx, p. 331, pl. ix, fig. 5.

Disc thin, membranous, seldom preserved; arms slender, one inch long; oral pentagon conspicuous, of a beautiful petaloid shape, constructed, according to Mr. Salter, of a series of ogives, the salient angles of which are inserted into the base of the arms and formed of three pairs of bones, while in *P. Miltoni* only two are distinct, and these are set in an angular form (fig. 17, *c*). Small spines are fixed to their extremities.

Locality.—Leintwardine in Lower Ludlow Rock.

Protaster Sedgwickii, Forbes. 'Mem. Geol. Surv.,' dec. i, pl. iv.

" The upper and under surfaces of the disc were covered by small, similar, more or less regular, polygonal or crescentic plates, imbricated in scale-fashion, and having punctated surfaces. Those of the under side of the body are smaller and more regular than those of the upper. The mouth is central, and rather small in proportion to the disc. The buccal apparatus is composed of ten parts or processes, arranged in pairs; half of each springs from the origin of each arm in a diverging manner, and meets the corresponding half to form a lanceolate, tooth-like projection, deeply indenting the cavity of the mouth. The arms were composed of alternating somewhat quadrate ossicula, the sides of which were deeply indentated superiorly, in order to form spiniferous crests. The spines were short, not equalling the length of an ossicle, obtuse, and few in a row; ambulacral grooves wide and convex; the central pentagon very deeply cut, of five oval, pointed petals.

Locality.—Underbarrow, Westmoreland, in Lower Ludlow Rock.

Protaster Salteri, Sow. 'Quart. Jour. Geol. Soc.,' vol. i, p. 20. 1845.

Discovered by Professor Sedgwick and Mr. Salter many years ago in Lower Silurian rocks, near Cerrig-y-Druidion on the Holyhead road; the original specimen has unfortunately been lost.

TÆNIASTER.—*Billings.*

This genus, according to Dr. Billings, differs from *Protaster* in the following particulars. *Protaster* has a well-developed disc, also pores outside of the ambulacral ossicles, and the oral plates are formed of two of the ambulacral ossicles ; whilst *Tæniaster* has no disc nor marginal plates, and the oral plates are formed of two of the adambulacral plates.

These important distinctions justify the separation of the species into a separate genus, which presents many affinities with the *Ophiuridæ*.

Tæniaster spinosus, Billings. 'Canadian Organic Remains,' dec. iii, pl. x, fig. 3, p. 81.

Description.—The specimens collected are about seven lines in diameter ; the rays linear-lanceolate, one line in width at the base, and covered at the sides with numerous small spines.

In the view of the enlarged specimen (pl. x. 36), the ambulacral ossicles appear in some places to alternate with each other, but this is owing to a distortion. Those on one side of the furrow are opposite those upon the other. The adambulacral plates are elongated, and so placed that the outer extremity of the one lies upon the inner extremity of the next. The rays are flexible.

Locality and Formation.—Falls of Montmorency, Trenton Limestone, collected by Dr. Billings.

Tæniaster cylindricus, Billings. 'Canadian Organic Remains,' dec. iii. pl. x. fig. 4., p. 81.

Description.—About an inch and a half in diameter, or a little more ; rays sub-cylindrical, regularly rounded on the upper side, flattened on the lower, covered above with spines ; about a line in width at the base, and tapering to an acute point.

This species is larger and more robust than the former. Both appear to be somewhat common, and the specimens are often found with their rays variously curved, showing that they were extremely flexible.

Locality and Formation.—City of Ottawa, Trenton Limestone. Collected by Dr. Billings.

PALÆODISCUS.—*Salter.*

This genus has been proposed by Mr. Salter for a unique Asteroid, found at Leintwardine. The disc, which is incomplete, is formed of large, irregular, rhomboidal plates, which radiate in seven or eight rows from the mouth in each inter-ambulacral space, and are furnished with short spines. The only specimen at present known is

Palæodiscus ferox, Salt. 'Ann. and Mag. Nat. Hist.,' 2d series, vol. xx, p. 333, Pl. IX, fig. 6.

"The inner angles of the inter-ambulacra (fig. 18) form a prominent triangular boss, cut off distinctly by a furrow, and between these are the large elongated basal plates of the ambulacra, lying parallel, and not at all divergent; they are thick and blunt, and together with the five bosses form a circumscribed star in the centre, the massive character of which contrasts strikingly with the thinness and delicacy of all the other parts either of the ambulacra or disc. The ambulacra are small, composed of a double row of transverse plates, narrow, and more crowded than in *Palasterina* or *Palæaster*, and apparently very thin in texture. They

FIG. 18.

Palæodiscus ferox, Salt.

can be detached and leave the upper plated surface free, which is covered with irregular plates. If there be no deception in this—for we have only a single specimen—the affinity would be much closer with the forms above mentioned, although the strong oral apparatus reminds us more nearly of *Protaster*.

Locality.—Leintwardine, in Lower Ludlow Rock; the only specimen is in the cabinet of Mr. Marston, of Ludlow.

Professor Forbes[1] described under the name *Lepidaster Grayi* a remarkable Echinoderm from the Wenlock limestone of Castle Hill, Dudley, which at first glance bears a resemblance to *Solaster*, a closer examination of its structure, however, shows that it possesses characters which justify its separation into a distinct genus, so widely different from other *Asteriadæ*, that it may possibly form a connecting link between that order and the *Crinoideæ*.

"The disc of Lepidaster," says Professor Forbes, "is very little more than two inches in diameter. It is, unfortunately, so much injured, that the elements cannot be clearly made

'Memoirs of the Geological Survey,' decade iii.

out, but appears to have had a framework composed of closely-set polygonal ossicula. Around it are arranged the rays, equidistant from each other, like so many spokes of a wheel. Their average length is one inch and one-twelfth, and their breadth towards the base, four-twelfths. They are regularly lanceolate. Their under surfaces are exposed on the slab, and are composed of thick, transversely oblong plates, slightly overlapping each other in scale-like fashion, and ranged in four longitudinal rows, two on each side of a central or ambulacral groove, which is itself towards the extremity, in some instances, partially filled up by small polygonal intervening plates. Of the two rows of border plates on each side of the groove, the inner series is formed of oblong obscurely hexagonal ones, with traces of punctations and grooves on the surfaces, as if for spines. The ray that is most perfect exhibits twenty-five plates in each row. The outer series consists of sub-orbicular or obscurely polygonal plates, which, like the inner ones, are gently convex on their surfaces. The upper surface of the ray, and probably of the body, was composed of numerous small, polygonal, nearly flat ossicula, closely set, and of various sizes."

This remarkable fossil was associated with several species of Crinoids, both perfect specimens and in fragments, but the most careful search has never brought to light another specimen of *Lepidaster*.

Besides the species above enumerated, other Palæozoic Asteroidea have been mentioned by the following authors.

Hisinger[1] described and figured *Palasterina antiqua* from Ludlow rocks at Mount Hoburg, Gothland, Sweden.

Professor James Hall[2] has figured and described *Palæaster matutina* from the Lower Silurian rocks of the State of New York.

Mr. Troost[3] has figured and described *Asterias antiqua* from Silurian rocks in the State of Tennessee, and the author alludes to five other undescribed species, from beds of the same formation.

Mr. Locke[4] has recorded the same species, *Asterias antiqua*, from the Lower Silurian rocks of the State of Pennsylvania.

Johannes Müller[5] has described as *Asterias Rhenana*, a star-fish with marginal plates, obtained from Devonian Sandstone at Coblenz.

M. Thorent[6] has figured and described a Palæozoic asteriad, under the name *Asterias constellata*, collected from Dumont's " Terrain anthraxifere," in France, in the department of l'Aisne ; this star-fish belongs to the same group of forms which Forbes,

[1] Hisinger, 'Lethea Suecica,' t. lxxxix, t. 26, fig. 6.

[2] Hall, 'Palæontology of New York,' vol. i, t. 29, fig. 5.

[3] Troost, 'Transact. Geological Soc. Pennsylvania,' vol. i, p. 232, t. 10, fig. 9.

[4] Locke, 'Proc. Ac. N. S. Phil.,' vol. iii.

[5] Müller, 'Verh. Naturh. ver. für Rheinl. und Westphl. Jahrg.,' xii, 1855.

[6] Thorent, 'Mémoires Soc. Geol. de France,' tome iii, tab. 22, fig. 7.

in his memoir on the Palæozoic Asteriadæ figured as *Uraster*, some of which are now grouped in the genus *Palæaster*.

M. Goldfuss[1] has figured as *Asterias obtusus*, a remarkable asteriad which was obtained by Alberti from the Muschelkalk of Friedrichshall in Würtemberg.

It has been long known to palæontologists that a remarkable change is observed in the generic characters of many fossil animals found in the secondary formations when compared with those belonging to the same classes which are entombed in the Palæozoic series. Reptiles, fishes, mollusca, and articulata afford abundant evidence of such change, and the sub-kingdom radiata supplies additional proofs of the same organic law.

As the Oolitic Asteriadæ will be figured and described in detail in this monograph, it is only necessary to state here that nearly all the species belong to the genera *Uraster*, *Astropecten*, *Luidia*, *Plumaster*, and *Goniaster*. The fossil species all appertain to extinct forms, but the genera to which they belong are nearly all living in our present seas.

The Cretaceous Asteriadæ, which have already been beautifully figured in Dixon's ' Geology of Sussex,' will form the subject of a future monograph, to succeed the Cretaceous Echinidæ now in course of publication. These interesting star-fishes of the Chalk period range themselves for the most part in the genera *Oreaster*, *Goniodiscus*, *Astrogonium*, *Stellaster*, and *Arthraster*.

The Tertiary formations have hitherto yielded very few species of ASTERIADÆ. Those from the English tertiaries, representing *Astropecten* and *Goniaster*, have been figured by Professor Edward Forbes,[2] in his Radiaria of the London Clay; and the specimens from the Vienna basin, likewise belonging to the genera *Astropecten* and *Goniaster*, have been figured and described by Dr. Camil Heller in his paper " über neue fossile Stelliriden."[3]

CLASSIFICATION OF THE ASTERIADÆ.

The numerous forms presented by the animals of this division early induced Llhwyd and Petiver to give generic names, as *Asteriscus* and *Echinaster*, to certain of the group. The first systematic monograph which appeared was published by J. H. Linck[4] of Leipsic; this work was illustrated by forty-two well-executed plates, representing the leading forms known to him. This author divided the Star-fishes (*Asterias*, Linn.) into two sections, I. STELLIS FISSIS.—II. STELLIS INTEGRIS, characterised by the presence or absence of ambulacra on the underside of the rays. In the first section, which

---

[1] Goldfuss, 'Petrefacta Germaniæ,' Band i, p. 208, t. 63, fig. 3.

[2] Palæontographical Society Volume for 1852.

[3] 'Akademie der Wissenschaften,' Band xxviii des Jahrgunges 1858, Wien.

[4] 'De Stellis Marinis liber singularis. Tabularum aenearum figuras,' Lepsic, 1733.

corresponds with our order Asteroidea, Linck proposed seven well established genera, as *Pentagonaster, Pentaceros, Astropecten, Palmipes, Stella coriacea, Sol marinus, Pentadactylosaster*, which have formed the basis of subsequent classifications. The "Class 1. Stellarum pauciorum quam quinque radiorum," comprising *Trisactis* and *Tetractis*, included forms which were either varieties or malformations of other species; and "Class III. Stellas marinas in plures quam quinque radios fissas," comprising *Hexactin, Heptactin, Octactis, Enneactin, Decactis, Dodecactis*, and *Triskaidecactis*, are in part included in the Sun-stars of modern authors.

De Blainville[1] divided the Asteriadæ into six sections; these nearly correspond with Linck's divisions, which were based on the form of body.

A. Species with a pentagonal body, and few or no lobes at the circumference, the angles being fissured (LES OREILLERS, *Asterias discoidea*, Lamk.). Ency. Méthod. Pl. 97, fig. 3. *Asterias granularis*, Linn. Linck, Stellis Marinis, tab. xiii, fig. 3.

B. Species pentagonal, body thin and membranous (LES PALMASTÈRIES = PALMIPES, Linck). *Palmipes membranaceus*, Retz. Forbes, Star-fishes p. 116. *P. rosaceus*, Lamk. Enc. Méth., Pl. 99, fig. 2.

C. Species five-lobed, and not articulated at the circumference; ex. *Asterias minuta*, Linn. Enc. Méth., Pl. 100, fig. 1—3. *Pentaceros gibbus plicatus*, Linck, Stellis Marinis, t. 3, No. 20.

D. Species pentagonal, more or less lobed and articulated at the circumference (LES SCATASTERIES; ou PLATASTERIES). Examples, *Pentagonaster semilunatus*, Linck, Stell. Mar. t. 23, fig. 37. *Goniaster equestris*, Gmelin, Forbes, Brit. Star-fishes, p. 125.

E. Species deeply divided into five rays.
   * Rays triangular, depressed, and articulated at the borders (*Astropecten*, Linck; *Crenaster*, Luid.); examples, *Astropecten aurantiacus*, Linn. Forbes, Brit. Star-fishes, p. 130. Linck, Stell. Mar. t. 5 and 6.
   ** Rays triangular, short, and rounded above; example, *Uraster rubens*, Linn. Forbes, Brit. Star-fishes, p. 83. *Uraster glacialis*, Linn.; Forbes, Brit. Star-fishes, p. 78.
   *** Rays long, narrow, and often contracted at their origin. *Linckia variolata*, Agass. Linck, Stell. Mar. tab. i, fig. 1, tab. viii, fig. 10.

F. Species which have more than five or six rays (LES SOLASTERIES). *Solaster papposa* Linn. Forbes, Brit. Star-fishes, p. 112, Stell. Mar., tab. xvii, No. 28, tab. xxxii, No. 52. *Luidia fragillissima*, Forbes, Brit. Star-fishes, p. 135.

[1] 'Dictionnaire des Sciences Naturelles,' art. "Zoophyte," p. 216, 1830.

Nardo[1] proposed to divide the European star-fishes into five genera, several of which correspond with the Linckian genera : 1st. STELLARIA = ASTROPECTEN, Linck (includes *A. aurantiaca, A. calcitrapa*). 2nd. STELLONIA = STELLA-CORIACEA, Linck (*A. rubens, A. glacialis*). 3rd. ASTERINA = PENTACEROS, Linck *(A. exigua, A. minuta)*. 4th. ANSEROPODA = PALMIPES, Linck *(A. membranacea, A. rosacea)*. 5th. LINCKIA = PENTADACTYLOSASTER, Linck *(A. lævigata, A. variolosa)*.

Professor Agassiz,[2] in his arrangement of the Echinodermata, proposed a division of the Asteriadæ into nine genera, in which he substituted new names for Nardo's genera, and added others for extra European and extinct species. This arrangement is irrespective of the number of the rays.

1. ASTERIAS, Ag. (= *Astropecten*, Linck ; *Crenaster*, Luid. ; *Pentastérie*, Blainv. ; *Stellaria*, Nardo.) The body stellate, the upper surface tesselate, and the rays depressed ; the margin bordered with two rows of large plates, carrying small spines. Types of this genus, *A. aurantiaca, A. calcitrapa*.

2. CÆLASTER, Ag.—This genus differs from the preceding in having the inner cavity circumscribed by plates disposed as in the *Echinidæ*, and having in the anal area an ambulacral star. Type, *C. Couloni*, Ag.

3. GONIASTER, Ag. (= *Pentagonaster*, Linck ; *Pentaceros*, Linck ; *Scutastérie*, Blainv. The body pentagonal, margin bordered by a double series of large plates which support spines and granules, upper surface nodulated. Types. *A. tessellata*, Lamk. ; *A. equestris*, Linn.

4. OPHIDIASTER, Ag.—Rays long, cylindrical, or conical, covered with fine, close-set granules, ambulacral area very narrow. Type, *Asterias ophidiana*, Lamk.

5. LINCKIA, Nardo (= *Pentadactylosaster*, Linck). Body stellate ; rays elongated, covered with tubercules, showing the porous integument in the intertubercular spaces. Type, *A. variolata*, Lamk. ; Linck, Stel. Mar., tab. viii, fig. 10.

6. STELLONIA, Nardo (= *Stella coriacea*, Linck ; *Pentastéries* in part and *Solastéries*, Blainv.) The body stellate and covered with spines more or less prominent. Types, *A. rubens*, Linn. ; *A. glacialis*, Linn. ; *A. endeca*, Linn., Linck, tab. xv. fig. 1 ; *A. papposa*, Linn., Linck, tab. xvii, No. 28 ; *A. helianthus*, Lamk. ; Enc. Méth., Pl. 108.

[1] 'Naturforscher,' 1833. 'Isis,' 1834.
[2] 'Mémoires Soc. Sciences Naturelles de Neufchatel,' tom. i, p. 190, 1836.

7. ASTERINA, Nardo (= *Asteriscus*, Luid.; *Pentaceros*, Linck). Body pentagonal, covered with pectinated scales, upper surface convex and inflated, under surface with deep, narrow, ambulacral areas. Type, *A. minuta*, Blainv., Linck, tab. iii, No. 20.

8. PALMIPES, Linck (= *Palmastérie*, Blainv.; *Anseropoda*, Nardo). Body pentagonal, much depressed, thin and membranous at the border. Type, *A. membranacea*, Retz., Linck, tab. i, fig. 2.

9. CULCITA, Ag. (= *Oreiller*, Blainv.) Body discoidal, pentagonal, fissured at the angles; the tegumentary membrane covered with granules. Type, *A. discoidea*, Lamk.; Encycl. Méthod., Pl. xcvii—xcix.

Dr. J. E. Gray published[1] a synopsis of the genera and species of Star-fishes in the collection of the British Museum and Zoological Society. The order ASTEROIDEA, or class HYPOSTOMA, of the author was thus subdivided :

SECT. 1.—*Ambulacra with four rows of feet; dorsal wart simple.*

Family 1.—ASTERIADÆ, including two genera.

1. ASTERIAS.—Skeleton netted with a single mobile spine at each anastomosis of the ossicula; body covered with more or less prominent, elongated, mobile spines. Type, *Asterias glacialis*, Linck, Stel. Mar. t. 38·39.

2. TONIA, Gr.—Skeleton netted with a series of crowded, small, blunt, mobile spines on the sides of each ossiculum; ambulacra bordered with a crowded series of subulate spines, and without any triangular pierced pieces within. *Tonia Atlantica*, Gr.

SECT. 2.—*Ambulacra with two rows of feet.*

Family 2.—ASTROPECTINIDÆ.

Back flattish, netted with numerous tubercles, crowned with radiating spines at the tip, called paxillæ.

1. NAURICIA, Gr.—Ambulacral spines broad and ciliated; two series of tesseræ between the angles of the arms and the mouth beneath. Asiatic. *N. pulchella*, Gr., Seba. iii, t. 8, fig. 7.

---

[1] 'Annals and Magazine of Natural History,' vol. vi, pp. 175 and 275, 1841.

2. ASTROPECTEN, Linck.—Ambulacral spines simple, linear, without any tesseræ between the marginal tubercles near the mouth and angles of the arms. *A. corniculatus*, Linck., Stel. Mar., t. 27 and 36.

3. LUIDIA, Forb.—Margin of the flat rays erect; dorsal surface crowded with regular paxillæ. *L. fragillissima*, Forbes, Brit. Star-fishes, p. 135.

4. PETALASTER, Gr.—Margin of the rays shelving; dorsal surface with equal paxillæ placed in longitudinal and transverse series. Asiatic, *P. Hardwickii*, Gr., Brit. Mus.

5. SOLASTER, Forb.—Rays many, with two series of broad spines bearing tubercles near the ambulacra. *S. papposa*, Forb., Brit. Star-fishes, p. 112.

6. HENRICIA, Gr.—Rays five, rounded, tapering, with rounded tubercles near the ambulacra; the dorsal wart obscure, concealed by spines. *H. oculata*, Penn.

Family 3.—PENTACEROTIDÆ.

The body supported by roundish or elongated pieces, covered with a smooth or granular skin, pierced with minute pores between the tubercles.

1. CULCITA, Ag.—Type, *Asterias Schmideliana*, Retz., Naturforscher., xvi, t. 1.

2. PENTACEROS, Linck.—Body convex above, margin with two rows of large spine-bearing tesseræ. *P. gibbus*, Linck., Stel. Mar., t. 23, fig. 36.

3. STELLASTER, Gr.—Body depressed, covered with large, flat, regular six-sided plates; margin with two rows of large tesseræ; the lower rows with a series of compressed mobile spines. *S. Childreni*, Gr., fig. 10.

4. COMPTONIA, Gr.—Body depressed, spinose? dorsal and oral disc covered with very small, flat plates; marginal ossicula large, without any mobile spines. *C. elegans*, Dixons, Fossils of Sussex, t. 22, fig. 9.

5. GYMNASTERIA, Gr.—Type, *G. spinosa*, Gr., Brit. Mus.

6. PAULIA, Gr.—Body five-rayed, formed of flat, granulated, spine-bearing, irregular ossicula on the disc and margin, without any two-lipped pores. *P. horrida*, Gr., Brit. Mus.

7. RANDASIA, Gr.—Body pentagonal, with a tubercular skin above, and large granular plates beneath and on the margin, without any two-lipped slits, but with one or

two small pores near the oral angle beneath, where the tubercles are rubbed off. Allied to *Culcita*. *R. Luzonica*, Gr., Brit. Mus.

8. ANTHENEA, Gr.—Body five-rayed, chaffy, with immersed, elongated tubercle-bearing ossicula; margin with regular rows of large tesseræ; both surfaces (especially the under) scattered with large two-lipped pores. *A. Chinensis*, Gr., Brit. Mus.

9. HOSIA, Gr.—Body five-rayed, formed of distinct, hexangular, nearly equal, slightly tubercular ossicula; back with small, and under surface with larger, two-lipped slits. *S. flavescens*, Gr.

10. HIPPASTERIA, Gr.—Body four or five sided, formed of roundish ossicula, with a large, truncated, central tubercle; upper and under surfaces with two-lipped pores. *H. Europæa*, Gr., var. of *Goniaster equestris*, Gmel.

11. CALLIASTER, Gr.—Body five-rayed, with flat immersed ossicula, armed with flat-based, deciduous, conical spines, and without any two-lipped slits on either surface. *C. Childreni*, Gr.

12. GONIASTER, Ag.—Ossicula flat, the dorsal ossicula granulated and armed with deciduous, flat-based spines; both surfaces destitute of two-lipped pores. *Goniaster cuspidatus*, Linck., Stel. Mar., t. 21 and 23, fig. 37.

13. PENTAGONASTER, Linck.—Body formed of convex, smooth, and spineless ossicula; ossicles of the under side with a sunk, central line, a central perforation and a small pit at each end. Marginal ossicula near the tips of the rays very large and swollen. *P. pulchellus*, Gr.

14. TOSIA, Gr.—Body convex, formed of smooth, spineless ossicula; dorsal and ventral ossicula entire, subequal, without any impressed line; marginal ossicula two-rowed, with a small intermediate one near each tip; dorsal wart triangular. *T. Australis*, Gr.

15. ECHINASTER, Pet.—Body star-like, granulated, depressed; back rather convex, with a circle of ten to fifteen conical warts! Ambulacral spines small, placed in groups, with a single continuous row of large slender spines near them. Spines very long, covered with a granular skin, and having generally a second articulation about one third the length of the base. *E. Ellisii*, Solander and Ellis., Zooph., t. 60—62.

16. OTHILIA, Gr.—Skin smooth, polished; ambulacra with two very close series of filiform spines. *O. spinosa*, Linck., Stel. Mar., t. 4, fig. 17.

17. METRODIRA, Gr.—Slightly granular; rays slender, with large single pores and small scattered spines on the back; smooth, and formed of regular, flat ossicula on the sides. *M. subulata*, Gr.

18. RHOPIA, Gr.—*Stellonia*, Ag. Ambulacral spines long, with several series of larger spines near them. *R. seposita*, Retz.

19. FERDINA, Gr.—Body flat; rays broad, convex and warty above, flat and uniform beneath; ambulacral spines short, united at the base. *F. flavescens*, Leach.

20. DACTYLOSASTER, Gr.—Rays cylindrical, nearly smooth, formed of regular oblong ossicula, each furnished with a central group of unequal, short, mobile tubercles; one dorsal wart. *D. cylindricus*, Lamk.

21. TAMARIA, Gr.—Rays cylindrical, formed of seven series of granular, convex, roundish ossicula, each of the upper ones with three or four unequal, and the lower ones with a central, short, blunt spine. *T. fusca*, Gr.

22. CISTINA, Gr.—Rays cylindrical, nearly smooth, formed of rows of three-lobed flat ossicula, each furnished with a central, mobile spine; one or two oblong dorsal warts. *C. Columbiæ*, Gr.

23. OPHIDIASTER, Ag.—Rays cylindrical, elongate, uniformly granular all over, without any spines; back with a small central group of larger tubercles; dorsal wart concave, with radiating or twisting grooves. *O. aurantius*.

24. LINCKIA, Nardo.—*L. typus*, Nar.; *P. miliaris*, Linck., Stel. Mar., t. 28, fig. 47.

25. FROMIA, Gr.—Rays from five to eight, flat, triangular, formed of flat-topped, granular tubercles. *F. millepora*, Lam., Seba. Thesaur., t. 8, *f. a. b.*

26. GOMOPHIA, Gr.—Rays elongate, cylindrical, tapering, with a terminal tubercle; back with large rounded tubercles; back of the rays with a series of large, conical, convex, tubercular spines; the spines near the ambulacra small, crowded. *G. Egyptiaca*, Gr.

27. NARDOA, Gr.—Rays cylindrical, spineless, formed of large, granular, convex ossicula. *P. variolatus*, Linck., Stel. Mar., t. 8, fig. 10.

28. NARCISSIA, Gr.—Body pyramidal, thin, coriaceous, uniformly granular; rays tapering, elongate, triangular at the base, formed of thin, flattened ossicula. *N. Teneriffæ*, Gr.

29. NECTRIA, Gr.—Body rather pyramidal, coriaceous, scattered with truncated warts, granular at the top; rays roundish, produced, edged with two series of flat, granular warts on each side, beneath largely granular. *N. oculifera*, Lamk.

30. NEPANTHIA, Gr.—Body small, flat; rays very long, cylindrical, tapering, not margined, formed, above and below, of many regular longitudinal and transverse series of flat-topped tubercles, furnished at the top with a series of elongate, spine-like granulations. *N. tessellata*, Gr., Brit. Mus.

31. MITHRODIA, Gr.—Rays cylindrical, elongate, spinulose; skeleton netted with small, scattered, rugose spines, and a series of large, clavate, spinulose spines, regularly articulated to a broad, expanded base on the sides of the arms. *P. reticulatus*, Linck., Stel. Mar., t. 6, figs. 10 and 16.

32. UNIOPHORA, Gr.—Body rather depressed; rays broad, blunt; skeleton formed of a series of transverse, oblong ossicula, each bearing a large, unequal-sized, sub-globular, articulated spine, placed in a longitudinal series; dorsal wart convex, complicated. *U. globifera*, Gr.

Family 4.—ASTERINIDÆ.

Body discoidal or pyramidal, sharp-edged; skeleton formed of flattish, imbricate plates; dorsal wart single, rarely double.

1. PALMIPES, Linck.—Body flat, thin, nearly membranous, margin radiately striated; the dorsal ossicula with a radiating tuft, and the oral ones with a transverse line of many thin mobile spines; ambulacral spines in oblique, rounded groups. *P. membranaceus*, Linck., Stel. Mar., t. 1, fig. 2.

2. PORANIA, Gr.—Body pyramidal, thick, five-rayed; skin above and below varnished, spineless; dorsal ossicula irregular; margin with two series of large ossicula, the lower ones produced, sharp-edged, and each furnished on the edge with a series of mobile spines. *Goniaster Templetoni*, Forbes, Brit. Star-fishes, p. 122.

3. ASTERINA, Nardo.—Body rather pyramidal, five-rayed; the back convex; the oral surface flat; the ossicula of each surface furnished with one or more mobile, tapering spines; the margin sharp-edged, each of the ossicula with a marginal series of spines; ambulacral spines placed in groups of four or five. *Asterina gibbosa*, Penn., Forbes, Brit. Star-fishes, p. 119.

4. PATIRIA, Gr.—Body pyramidal, coriaceous, five-rayed ; ossicula of the oral surface with uniform radiating groups of small spines ; those of the dorsal surface of two kinds, the one crescent-shaped, with series of small bundles of spines, the others bearing irregular round bundles of spines between them. *P. coccinea*, Gr.

5. SOCOMIA, Gr.—Body depressed ; rays elongate, formed of imbricate plates ; the margins broad, the upper and lower series of ossicles separated by a groove. *S. paradoxa*, Gr.

Müller and Troschel proposed a new classification of fifty-five species of Asteriadæ contained in the Berlin Museum,[1] this was afterwards amended by the introduction of other genera ; these memoirs formed the Prodrome of their ' System der Asteriden,' [2] which now constitutes a standard work upon the ASTERIADÆ. As this important monograph is not much known in England, no apology is necessary for giving the following translation of the synopsis of the families and genera contained therein.

SUMMARY OF THE FAMILIES AND GENERA OF THE ASTERIADÆ.

FIRST FAMILY.

Four rows of tentacula in the ambulacral avenues. With a vent.

1 *Genus.*—ASTERACANTHION, M. and T. = *Stellonia*, Nardo ; *Uraster*, Ag.

Rays long, many rows of spines near the ambulacral areas on the under side ; the whole of the upper surface of the body and rays covered with blunt or pointed spines, either scattered singly or grouped toge- ther into tufts, and arranged more or less regularly in rows. The integument be- tween the spines naked, showing the basis of the spines. In the naked skin many tentacule-pores. Pincers-like pedi- cellariæ, supported on soft stems, either encircling the basis of the spines or scattered between them, or both ; some likewise on the borders of the avenues. Vent subcentral.

FIG. 20.

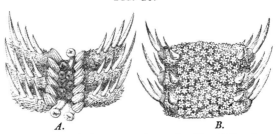

A.   *B.*

Portion of a ray of *Astropecten polyacanthus*, M. and T. *A*, under surface ; *B*, the upper surface of the ray.

---

[1] 'Monatsbericht der Königl. Akad. der Wissenschaft,' Monat, April, 1840.
[2] 'System der Asteriden mit Zwölf Kupfertafeln, Braunschweig,' 1842.

SECOND FAMILY.

Two rows of tentacula in the ambulacral avenues.   With a vent.

2 *Genus.*—ECHINASTER, M. and T. = *Pentadactylosaster*, Linck.

Rays long, conical, or cylindrical; the skin supported on a network of calcareous pieces, from which longer or shorter spines proceed, sometimes alone or set close together.   Skin between the spines naked, with many tentacule-pores.   Vent subcentral.

3. *Genus.*—SOLASTER, Forbes.

FIG. 21.

A.   *B.*

Portion of a ray of *Solaster papposa*, Linn.   *A*, the under; *B*, the upper surface.

Body stellate, multiradiate; rays moderately long, and covered with fasciculated spines; skin between the spines naked, with many pores; no pedicellariæ; vent central.

4. *Genus.*—CHÆTASTER, M. and T.

Rays narrow and elongated, covered with plates which carry on their summits fasciculated spines, single pores between the plates; vent subcentral.

5. *Genus.*—OPHIDIASTER, Agass. = *Linckia*, Nardo (pars).

FIG. 22.

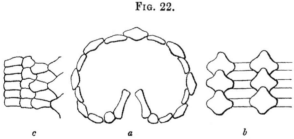

c   *a*   *b*

Section of a ray of *Ophidiaster*, with the inter-ambulacral plates.

Rays cylindrical or conical, everywhere covered with granular plates; many pores between the plates, which are likewise surrounded by granules; no pedicellariæ; vent central.

6. *Genus.*—SCYTASTER, M. and T. = *Pentadactylosaster*, Linck.; *Linckia*, Nardo.

Rays elongated; body stellate, everywhere covered with granular plates, which are arranged on the margins in two rows; the space between the plates likewise granulated. The pores between the plates single.   No pedicellariæ; vent subcentral.

7. *Genus.*—Culcita, Agass.

Body pentagonal, with a thick, blunt border, which forms a very high side-area, without plates at the margin. Body covered with granules, and fissured at the angles ; valve-like and pincers-like pedicellariæ ; vent subcentral.

8. *Genus.*—Asteriscus, Luid. = *Asterina,* Nardo ; *Palmipes,* Linck.

Body pentagonal, with short rays ; under surface flat, upper surface more or less inflated or completely depressed, thin and membranous at the border, without plates. Ossicula on the under side covered with small, pointed, or blunt, or cylindrical spines, which stand either simply on each plate, or comb-like in rows. The plates on the upper surface covered with similar processes, either comb-like or fasciculated ; single pores between the plates of the upper surface of the disc and of the rays. The pores cease, sooner or later, near the margin ; vent subcentral.

9. *Genus.*—Pteraster, M. and T.

In form it resembles the genus *Asteriscus,* with two rows of tentacula in the ambulacral avenues, and a central vent. The upper surface of the disc and rays is completely covered with a naked skin, from which tufts of small spines are regularly developed, and supported upon the osseous network of the ray. The border is armed with a series of longer spines, united together up to their extremities by a fold of the tegumentary membrane. The under side

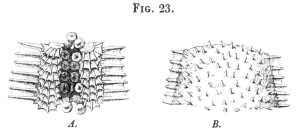

Fig. 23.

A portion of a ray of *Pteraster militaris,* M. and T.
A, the under ; *B,* the upper surface.

of the ray is provided with a series of transverse spines, formed into fan-like structures by folds of the skin, and superimposed on each other along the sides of the avenues ; there are no pedicellariæ.

10. *Genus.*—OREASTER, M. and T. = *Pentaceros*, Linck.; *Goniaster*, Agass. (pars).

FIG. 24.

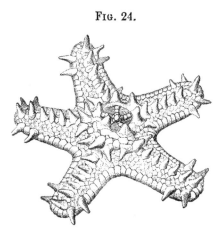

Upper surface of *Oreaster turritus*, Linck.

Under side flat, upper surface more or less elevated; the middle of the rays ridged with an angular, or strongly inflated, keel. Two rows of granular marginal plates; the border is formed by the upper row, and the lower row lies at the under surface. The body is everywhere covered with smaller or larger plates, which, with the marginal plates, are either entirely granular, or sometimes carry more or less spinous tubercles; the pore-fields between the plates of the upper surface are granulated with many pores. Pedicellariæ sessile, either valve-like or pincers-like; vent subcentral.

11. *Genus.*—ASTROGONIUM, M. and T. = *Pentagonaster*, Linck.; *Goniaster*, Agass.

Body pentagonal, disc-shaped, flat on both sides. Two rows of marginal plates, which are much larger than the other plates of the body. Both rows contribute to the formation of the margin. Their border is surrounded by a wreath of granules, or their circumference is encircled by granules. Up to this enclosure they are completely naked. Sometimes

FIG. 25.   FIG. 26.

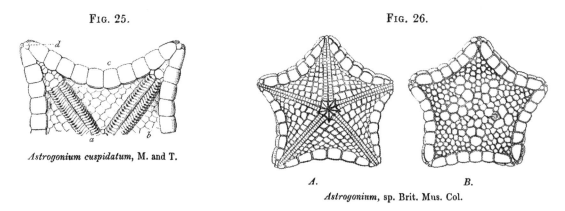

Astrogonium cuspidatum, M. and T.

A.   B.

Astrogonium, sp. Brit. Mus. Col.

they carry tubercles upon the middle. The upper and under surfaces are tesselated with free-lying plates; between these only narrow, granulated pores have places; vent subcentral.

12. *Genus.*—Goniodiscus, M. and T.

Body pentagonal, disciform, flat on both sides. Two rows of large marginal plates at the border, the whole upper surface of which is granulated. The upper as well as the under marginal plates contribute their share to form the thick border, and divide the same into two equal parts; the pentagonal form of the disc is therefore maintained by this double row of marginal plates on the flanks and border. The upper and under surfaces are tesselated with a different form of granulation; vent subcentral.

13. *Genus.*—Stellaster, Gray.

Fig. 27.

Body nearly pentagonal, flat on both sides, with two rows of large granulated marginal plates, both of which contribute to the formation of the high border. Each ventral marginal plate carries a suspended spine. Both surfaces of the disc are covered with granulated plates; vent subcentral.

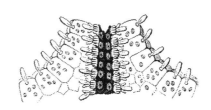

Under surface of a ray of *Stellaster Childreni*, Gray.

14. *Genus.*—Asteropsis, M. and T.

Rays short in proportion to the pentagonal form of the body; under surface flat; upper surface more or less elevated; the rays sometimes ridged. Two rows of marginal plates; the border sharp, and formed only of one row of plates. The integument on the upper and under surface contains ossicula, but the interspaces of the plates, sometimes also the plates themselves, and always the pore-fields of the upper surface, are completely naked; vent subcentral.

15. *Genus.*—Archaster, M. and T.

Body flat, with elongated rays. Two rows of large marginal plates; those of the lower series reach from the under part of the rays to the furrow-plates and are covered with scales, which at the border are metamorphosed into moveable spines; the upper series are covered with granules, which become bristle-like. The upper surface is level, and closely covered with appendages whose summits are crowned with small bristles. Between the paxillæ are single pores; vent central.

Third Family.

Two rows of ambulacral suckers in the avenues; without a vent.

16. *Genus.*—Astropecten, Linck.= *Stellaria*, Nardo; *Asterias*, Agass.

Body stellate, flat on both sides, with elongate rays. Two rows of large marginal

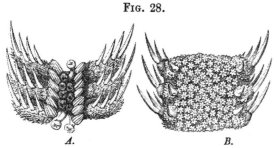

Fig. 28.

plates at the border. The lower series provided with spine-like scales, which increase from within outwards into long moveable spines. The dorsal marginal plates are covered with granules, which often become spinous, and sometimes carry spines. The flat upper surface of the body and rays thickly covered with appendages, the summits of which are

Portion of a ray of *Astropecten polyacanthus*, M. and T.   *A*, under surface; *B*, the upper surface of the ray.

crowned with groups of minute spines, as in *Archaster*.

17. *Genus.*—Ctenodiscus, M. and T.

Body flat, almost pentagonal; at the border two rows of marginal plates, which are completely smooth on the upper surface. The marginal plates on the under surface form transverse bands, which are clothed with scales. A comb-like ridge of fine spines surrounds the border of these plates. The upper marginal plates have a row of large spines at the border, and the basal marginal plates support a row of similar spines near the avenues. The dorsal surface is covered with paxillæ, as in *Astropecten*.

18. *Genus.*—Luidia, Forbes.

Fig. 29.

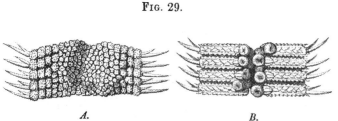

A.   *B.*

Portion of a ray of *Luidia Senegalensis*. M. and T.   *A*, the upper; *B*, the under surface.

Rays elongate. Instead of a double row of marginal plates, only a single row on the under side, with spines. The entire of the dorsal surface covered with paxillæ, as in *Astropecten* and *Ctenodiscus*.

DESCRIPTION

OF THE

FOSSIL OOLITIC ASTERIADÆ.

ON THE ASTERIADÆ OF THE LIAS.

BEFORE describing the Asteriadæ of the Lias I purpose giving a short account of the several zones of life into which this great formation is now divided, with the view of accurately defining the statigraphical distribution of its Echinodermata in time and space. More ample details on the subject than my present limits permit will be found in the various works cited in the synonyms prefixed to the description of the different zones.

English geologists divide this formation into Upper Lias, Marlstone, and Lower Lias, but these subdivisions require additions and modifications in order to place the liassic beds of the British Isles in correct correlation with those of France, Switzerland, and Germany. For on the Upper Lias clays, in certain localities, are superimposed extensive arenaceous deposits, which, previous to the publication of my memoir on the Upper Lias sands,[1] were grouped with the Inferior Oolite, and in the Lower Lias are included several beds of clays and marls which, with the Marlstone of English authors, forms the Middle Lias of continental geologists.

Taking the Lias beds so well exposed in their natural order of superposition in the north and south of England in the magnificent sections on the Yorkshire and Dorsetshire coasts, and naming each group of beds by the most characteristic Ammonite contained therein, we find the following zones of life, taken in descending order :

THE UPPER LIAS.—The sands of the Upper Lias forming the upper portion of this zone, are characterised for the most part by Ammonites belonging to the group *Falciferi*, as *Ammonites opalinus*, Rein., and *A. radians*, Schloth. ; *Ammonites Jurensis*, Ziet., and *A. insignis*, Schübl., both belonging to other groups, are likewise associated with them.

---

[1] PALÆONTOGRAPHICAL AND STRATIGRAPHICAL RELATIONS of the so-called " SANDS of the INFERIOR OOLITE," ' Quart. Jour. of the Geological Society,' vol. xii, p. 292, 1856.

The clays of the Upper Lias forming the lower part of the zone, are everywhere distinguished by other species of *Falciferi*, as *Ammonites bifrons*, Brug.; *A. serpentinus*, Schloth., and numbers of the group *Planulati*, as *Ammonites communis*, Sow.; *A. anguinus*, Rein., and *A. fibulatus*, Sow.

The Middle Lias.—This is divisible into five zones, each characterised in descending order by—*Ammonites spinatus*, Brug. 2. *A. margaritatus*, Mont. 3. *A. capricornus*, Schloth. 4. *A. Ibex*, Quenst. 5. *A. Jamesoni*, Sow.

The Lower Lias is divisible into seven zones. These are—1. Zone of *A. raricostatus*, Ziet. 2. *A. oxynotus*, Quenst. 3. *A. obtusus*, Sow. 4. *A. Turneri*, Sow. 5. *A. Bucklandi*, Sow. 6. *A. angulatus*, Schloth. 7. *A. planorbis*, Sow.

Complicated as these subdivisions may at first sight appear to those who have regarded the Lias merely as a great clay deposit, with a uniform fauna throughout, still their accuracy may be clearly demonstrated in the grand section on the Dorsetshire coast, extending from near Bridport harbour on the east, to Pinhay bay on the west. Within these limits the entire series of beds rise beneath each other on the shore, and are exposed in the cliffs, so that this coast section may be said to be complete from the great arenaceous deposit of Upper Lias sand, containing *Ammonites opalinus*, with each succeeding zone of the Upper, Middle, and Lower Lias, down to *Ammonites planorbis*, and its Ostrea series resting on the *Avicula contorta* beds of the Trias formation.

In the following table I have placed the different zones of the English Lias in correlation with those of Germany, so well described by Professors Quenstedt, Oppel, Fraas, and others; those of France by the late M. A. d'Orbigny, and of those of England by Sir R. Murchison, Sir H. De la Beche, and the Rev. W. D. Conybeare.

A TABLE SHOWING THE CORRELATION OF THE LIAS BEDS OF FRANCE, GERMANY, AND ENGLAND.

SWABIA. Quenstedt.	WURTEMBURG. Oppel.	FRANCE. D'Orbigny.	SOUTH OF ENGLAND. Wright.	GLOSTERSHIRE. Murchison.	DORSETSHIRE. De la Beche.
Lias ζ. Jurensismergel.	Jurensis-bett.	Toarcien.	Zone of AMMONITES JURENSIS.	Inferior Oolite.	Inferior Oolite. Lower part.
Lias ε. Posidonienschiefer.	Posidonomyen-bett.	UPPER LIAS.	Zone of AMMONITES COMMUNIS.	Upper Lias.	
Lias δ. Amaltheenthone	Spinatus-bett.		Zone of AMMONITES SPINATUS.	Marlstone.	Micaceous Marl.
	Margaritatus-bett.	Liasien.	Zone of AMMONITES MARGARITATUS.		
	Davæi-bett.		Zone of AMMONITES CAPRICORNUS.	Ochraceous Lias.	
Lias γ. Numismalismergel	Ibex-bett.	MIDDLE LIAS.	Zone of AMMONITES IBEX.		Belemnite Bed. } Upper Marl.
	Jamesoni-bett.		Zone of AMMONITES JAMESONI.	Hippopodium & Cardinia Beds. } Lower Lias Shales.	
Lias β. Turnerithone.	Raricostatus-bett.		Zone of AMMONITES RARICOSTATUS.		
	Oxynotus-bett.	Sinémurien.	Zone of AMMONITES OXYNOTUS.	Ammonite Bed.	Ammonites in nodules.
	Obtusus-bett.		Zone of AMMONITES OBTUSUS.		
	Tuberculatus-bett.	LOWER LIAS.	Zone of AMMONITES TURNERI.	Plagiostoma Beds.	Lias Limestones } Lower Lias.
Lias α. Sand u. Thonkalke.	Bucklandi-bett.		Zone of AMMONITES BUCKLANDI.	Lower Lias Limestones and Shales.	
	Angulatus-bett.		Zone of AMMONITES ANGULATUS.	Saurian and Insect Beds.	White Lias, upper part. } Limestones.
	Planorbis-bett.		Zone of AMMONITES PLANORBIS.		White Lias, inferior part.
Vorläufer des Lias.	Bone-bed.		Zone of AVICULA CONTORTA.	Black Shales and Bone Bed.	Lower Marl, Bone Bed. } White Lias.

The above table clearly shows how completely the whole Lias formation is represented in England, and how nearly it corresponds bed for bed with the Wurtemberg series. I shall now describe the different zones in ascending order, commencing with the lowest, the zone of *Ammonites planorbis*.

THE LOWER LIAS.

1. THE ZONE OF AMMONITES PLANORBIS.

Synonyms.—" White Lias," William Smith, 'Memoir to the Map,' p. 47, 1815. "White Lias (pars)," De la Beche, 'Geol. Trans.,' 2nd series, vol. ii, p. 26. "Saurian Beds," Murchison's 'Geology of Cheltenham,' 2nd ed., by Buckman and Strickland, p. 49, 1845. "Psilonotenbank," Quenstedt, 'Der Jura,' Table, p. 293, 1857. "Die Schichten des Ammonites planorbis," Oppel, 'Juraformation.' p. 24, 1856. "Zone of Ammonites planorbis," Wright, 'Quart. Jour. Geol. Soc.,' vol. xvi, p. 389.

This division of the Lower Lias is well developed in the South of England.[1] In general it consists of a series of thin, grayish, or bluish, argillaceous limestones, with alternating beds of laminated shale; or sometimes it forms the upper part of the thick-bedded argillaceous, cream-coloured limestone, called "White Lias" by William Smith. In the upper half of this group of beds *Ammonites planorbis*, Sow., in some localities is found in considerable numbers, compressed in the shales, with its white shell more or less preserved; in the lower portion of the series *Ostrea liassica*, Strickl.,[2] appears in great numbers; beneath these strata are three or four beds of hard limestones (or "fire-stones"), in which the finest skeletons of *Enaliosauria* have been discovered. As this distinction, by means of *Am. planorbis, Ostrea liassica*, and Saurians, is a practical and useful one in the investigation of this zone of life, I shall adhere to it on the present occasion,—premising, however, that *Ammonites* are very rare in the lower beds, although abundant in the upper; and that *Ostreæ* are abundant below, but rare above, whilst Saurian remains are found throughout the entire series.

The best sections of the zone of *Ammonites planorbis* are those afforded by the extensive quarries at Street and the coast section at Watchet, in Somerset; at Binton

---

[1] The substance of the following observations on the Lower Lias is contained in my memoir on the "ZONE OF AVICULA CONTORTA, and the LOWER LIAS of the SOUTH OF ENGLAND," 'Quart. Jour. of the Geological Society,' vol. xvi, p. 374, 1860.

[2] *Ostrea liassica*, Strickland, is a very characteristic shell of the lowest Lias beds. It resembles *Ostrea irregularis*, Münster (Goldfuss, 'Petr. Germ.,' pl. vii—ix, fig. 5), and *Ostrea rugata*, Quenstedt ('Jura,' pl. iii, fig. 17). Dunker, in the 'Palæontographica,' (pl. vi. fig. 27), has figured a small Oyster from the Lias of Halberstadt (*Ostrea sublamellosa*, Dunker), which appears to be identical with our species. These small, thin, rugose Oysters are found in great abundance in the lowest beds of the Lower Lias in England and Germany. They are probably only varieties of one species, which had a wide geographical distribution in the seas which deposited the basement-beds of the Lias.

and Wilmcote, in Warwickshire; and at Pinhay Bay and Up Lyme, in Dorset. I purpose giving a detailed description of the most typical sections in each county.

Somersetshire.—At Street the strata are nearly horizontal and undisturbed, and therefore admit of accurate measurement. The following section of Mr. Cree's quarry I have compared with like sections afforded by the quarries of Messrs. Seymour, Underwood, and Talbot in the same parish; and find that the variation is so inconsiderable that any one may be said to represent the others, both as regards the sequence of the beds and the fossils they contain.

Section of the Zone of Ammonites planorbis, *at Street, Somerset.*

No.	LITHOLOGY.	ft.	in.	ORGANIC REMAINS; AND THE LOCAL NAMES OF BEDS.
				PLANORBIS SERIES.
1.	Light-coloured marly clay	3	0	"Top Bed." Saurian bones and *Ammonites planorbis.*
2.	Light-coloured Lias limestone	0	9	*Ammonites planorbis* in moulds.
3.	Yellowish laminated shale, splitting up into thin layers	3	0	*Ichthyosaurus intermedius, Ammonites planorbis, Lima punctata,* and *Isastræa Murchisoni.*
4.	Light-coloured shaly limestone	0	4	*Ammonites planorbis,* compressed.
5.	Hard gray limestone	0	7	"Building-stone." *Ammonites planorbis,* on the top of the bed, *Lima punctata* and *Lima gigantea.*
6.	Dark-gray shale	0	3	*Ammonites planorbis,* and muricated spines of *Cidaris, Edwardsi.*
7.	Dark-gray limestone	0	6	"Corn-sized building-stone." Spines of *Cidaris* and bones of *Ichthyosaurus tenuirostris.*
				OSTREA SERIES.
8.	Dark laminated shale	0	4	*Ostrea liassica.*
9.	Dark-gray limestone	0	5	"Five-inch building-stone." *Ostrea liassica.*
10.	Dark shale	0	3	*Ostrea liassica.*
11.	Dark-gray limestone	0	6	"Six-inch building-stone." *Cardinia crassiuscula, Lima punctata,* and *Ostrea liassica.*
12.	Dark shale	0	6	
13.	Grayish hard limestone, consisting of two 4-inch beds	0	8	"The Whetstones." Best paving-bed. Fossils rare: *Ostrea liassica* and *Modiola minima.*
14.	Hard dark marl	0	9	"Saurian bed." Many Saurians have been obtained here. *Ichthyosaurus intermedius* and *Plesiosaurus Hawkinsii* (British Museum). Jaws of Saurians and Fishes.
15.	Grayish fine-grained limestone	0	3	"The Cream-bed." Fine-grained paving-stone. *Ostrea* and *Modiola.*
16.	Brownish limestone	0	4	"Red Liver." Paving-stone. Few fossils.
17.	Dark-coloured limestone	0	4	The "Black stone," used for large paving-slabs; some of them 10 ft. by 5 ft. *Modiola minima, Ostrea liassica, Myacites,* and *Rhynchonella variabilis.*

No.	LITHOLOGY.	ft.	in.	ORGANIC REMAINS; AND THE LOCAL NAMES OF BEDS.
18.	Dark-blue shale	0	2	*Ostrea liassica* and *Modiola minima.*
19.	Hard grayish limestone	0	6	"Six-inch building-stone." *Ceromya, Modiola minima,* and *Ostrea liassica.*
20.	Soft bluish shale	0	2	
21.	Grayish-blue limestone	0	4	"Four-inch building-stone." Fossils as in No. 19.
22.	Dark-gray laminated shale	0	4	*Ichthyosaurus intermedius* and *I. tenuirostris.*
23.	Hard blue limestone	1	0	"The Blue Clog," or "One-foot building-stone," used for steps. *Ceromya, Ostrea, Modiola,* and *Rhynchonella.*
24.	Gray laminated shale	1	3	Saurians abundant: *Ichthyosaurus intermedius* and *I. tenuirostris, Pholidophorus leptocephalus,* Agass.
25.	Grayish limestone	1	0	"Gray Clog." A valuable building-stone, used for steps, troughs, &c. *Modiola minima.*
26.	Dark shale	0	2	
27.	Thin-bedded limestone	0	3	"Three-inch blue bed." Fish-remains, *Modiola minima,* and *Otopteris acuminata,* L. & H.
28.	Thick blue limestone	0	5	
29.	Hard fine-grained limestone	0	4	"Fire-stone."
30.	Hard, gray, fine-grained limestone	0	4	*Plesiosaurus Etheridgii.* (In the Jermyn Street Museum; and another now in Street are from this bed.)
31.	Hard gray limestone, forming the bottom bed	1	0	"Fire-stone, bottom bed." *Plesiosaurus Hawkinsii.* [The large *Ples. megacephalus,* Stutch., now in the Bristol Institution, was obtained from this bed near Street.]

The Saurian beds near Langport have likewise yielded Reptilian remains. I have obtained two fine specimens of *Ichthyosaurus intermedius* and an imperfect specimen of *I. tenuirostris* from this locality, which are now in the collections of private friends. In connexion with these Saurian beds of Somerset, it is important to note that the oldest *Enaliosauria* of the Lias are *Plesiosauri; Plesiosaurus Hawkinsii,* Owen, and *Pl. Etheridgii,* Huxley, were both exhumed from the 4-inch firestone, forming the bottom bed of the Ostrea series; the remarkable *Plesiosaurus megacephalus,* Stutch., now in the Bristol Museum, was found likewise in the firestones of a quarry near Street Foss, and it will be shown in my section of the correlative beds of this zone at Wilmcote, in Warwickshire, that the fine skeleton of *Plesiosaurus megacephalus* contained in the Warwick Museum was likewise exhumed from the "firestones" of that locality—beds which are the precise equivalents of the "firestones" of Street.

The small number of *Conchifera* hitherto found in these beds is very remarkable. *Ostrea liassica,* Strickl., *O. irregularis,* Quenst., *Modiola minima,* Sow., *M. psilonoti,* Quenst., *Gervillia,* n. sp., *Anomya,* n. sp., *Myacites,* n. sp., *Arca,* n. sp., and *Cardium,* n. sp., are the only species I have as yet collected from the firestone-beds.

This section likewise settles a point which has been often discussed, namely, what is the age of the Saurian beds of Somerset? It has been generally supposed that they appertain to the same horizon as the lower Saurian beds at Lyme Regis; this, however, is a mistake, inasmuch as the Saurian beds at Street belong to the zone of *Ammonites planorbis*, whilst the Saurian remains at Lyme Regis, on the contrary, are for the most part found in and above the zone of *Ammonites Bucklandi*.

Worcestershire and Gloucestershire.—The *Am. planorbis*, *Ostrea*, and lower Saurian beds, so well developed at Street, are likewise found in different parts of Worcestershire and Gloucestershire, where they present the same stratigraphical relations, and yield the same organic remains.

The neighbourhood of Tewkesbury affords several good exposures of the infra-ammonite beds. I have obtained *Ichthyosaurus tenuirostris*, Conyb., and *Ichthyosaurus intermedius*, Conyb., from a bed of light-coloured Lias at Haselgrove, near "the Folly;" and the late Mr. Dudfield, of Tewkesbury, collected several very fine skeletons of *Ichthyosaurus tenui-rostris*, Conyb., *I. intermedius*, Conyb., and *I. communis* (?), Conyb., with bones of *Plesiosaurus Hawkinsii*, at Brockeridge Common, and from similar beds at other localities around Tewkesbury; and I possess several vertebræ of *Plesiosaurus rugosus*, Owen, which were obtained from a bed of White Lias at Woolridge, near Hartpury.

The junction of the Lower Lias with the red marls of the Keuper in the Vale of Gloucester is, in general, indicated by a low escarpment, facing the west. At Brockeridge and Defford Commons this natural boundary is exceedingly well marked, and between these two localities are several quarries which expose to a greater or less extent the beds now under consideration. The presence of *Ammonites planorbis* in the upper strata of several of these sections has enabled me to correlate the beds beneath with the corresponding strata at Street, in Somerset, and at Binton, Grafton, and Wilmcote, in Warwickshire.

Section of the Ammonites planorbis, Ostrea, *and* Lower Saurian beds *at Brockeridge and Defford Commons.*

Zones.	Strata and Organic Remains.	Brockeridge.		Strensham.	
		ft.	in.	ft.	in.
Ammonites planorbis Beds.	1. Light-coloured clay	3	0	3	0
	2. White laminated limestone. " First rub," Brockeridge ; " Chance rub," Strensham..................................	0	4	0	4
	3. Brown laminated clay, with compressed white shells of *Ammonites planorbis*	3	0	2	0
	4. Blue argillaceous limestone ⎫ " Double rub," Brockeridge ;	0	3	0	2
	5. Brown shaly clay........... ⎬	0	2	0	2
	6. Blue limestone.............. ⎭ " Double nurf," Strensham.	0	3	0	2
Ostrea and Lower Saurian Beds.	7. Dark clay, with Saurian remains. " Yard clay "..................	3	0	3	0
	8. Hard blue limestone. *Ostrea liassica* on the surface of the rock. This bed is called " Red nurf " at Brockeridge, " King's nurf " at Strensham	1	0	0	3
	9. Dark clay. The second bed of " Yard clay " at Strensham	1	6	3	0
	10. Blue limestone. The " Queen's nurf," Strensham	0	3	0	3
	11. Blue clay	0	0	0	3
	12. Hard blue limestone, with *Modiola minima*	0	0	0	6
	13. Paving-stone, separated by an inch-band of clay	0	0	0	4
	14. Dark shale. Vertebræ of *Ichthyosaurus*, tests and spines of *Cidaris Edwardsi*, *Hemipedina*, sp., and Fishes' scales	0	0	0	6
	15. Hard blue limestone, in square blocks. " Brick-bed"...............	0	0	0	5
	16. Dark shale	0	0	0	3
	17. Insect limestone ; a hard argillaceous limestone, containing the Elytra and other remains of Insects	0	0	0	6
	18. Blue shale...	0	0	1	3
	19. Light-blue limestone, with *Cardinia*, sp., *Arca*, sp., and *Astarte*, sp.	0	0	0	4

I have placed the above sections together for the purpose of comparison : they were first made by my friend, the Rev. P. B. Brodie, and have been subsequently examined by myself with similar results. These sections show the uniformity which prevails in the Lower Saurian beds of Gloucestershire and Worcestershire, and how much they resemble their correlative strata at Street.

The late Mr. James Dudfield, of Tewkesbury, obtained from the infra-ammonite Lias beds at Brockeridge enumerated in the preceding sections, and from other strata occupying the same horizon in the vicinity of that town, a very fine series of Saurian remains, which were all sold and dispersed in June, 1843. From my notes of that collection I find there was a specimen of *Ichthyosaurus intermedius*, about 8 feet in length; the two fore-paddles and a portion of the scapular arch were tolerably complete; and there were upwards of 100 vertebræ and ribs nearly all in place. *I. tenuirostris;* 4 feet in length; the skulls, jaws, and teeth well preserved, the vertebral column tolerably complete; and likewise one fore-paddle. *I. communis;* very fine paddles. *I. platyodon;* large skull, with orbital plates in position. *Plesiosaurus Hawkinsii;* the vertebral column, ribs, and humeri; and fifty vertebræ in position.

The *Ostrea* and lower Saurian beds at Binton, Brockeridge, and Street are overlaid by clays and laminated shales, containing *Ammonites planorbis.* As these beds form a most important horizon in the Lias formation, and have a wide geographical distribution in England, France, and Germany, they require to be defined with accuracy, especially as some authors are of opinion that the true Lias commences with this zone of life.

The relation of the *Am. planorbis* shales to the Saurian beds below is extremely well exposed in the Railway-cutting at Uphill and in the quarries at Street, Binton, and Wilmcote, in Warwickshire, at Brockeridge Common, in Gloucestershire, and at Strensham, Worcestershire, and to the *Am. Bucklandi* or *Lima* beds above in the sections at Saltford, near Bristol; Penarth Head, near Cardiff; and Pinhay Bay, near Lyme Regis.

The following section of the beds at Binton was made by Mr. Robt. Tomes, of Welford Hill, near Stratford-on-Avon, from a quarry now abandoned. A similar exposition, however, is seen in the quarry worked near the former, the various beds of which I have examined and measured with Messrs. Tomes and Kershaw.

Section of the Zones of Ammonites planorbis *and* Avicula contorta, *at Binton, Warwickshire.*

No.	LITHOLOGY.	Thickness. ft. in.		ORGANIC REMAINS; AND LOCAL NAMES OF THE BEDS.
1.	Light-coloured limestone	0	6	" Top rock " or " Whites."
2.	Light-coloured clay	2	6	
3.	Argillaceous limestone	0	3	" Top Liveries." *Ichthyosaurus;* on the upper surface; Insects.
4.	Light-coloured clay	7	0	
5.	Argillaceous limestone	0	3½	" Top Liveries" (lower). Insects; *Ammonites Johnstoni,* Sow.
6.	Clay	1	1	
7.	Grayish limestone	0	6	" Extra rock." " Thick paving-bed." No fossils.
8.	Clay	0	3½	
9.	Grayish limestone. Thin and irregular when covered by the preceding 2 in. to	0	3	" Quarters."
10.	Clay	0	8½	
11.	Grayish limestone. A constant bed .	0	3½	" Ribs." Insects.
12.	Clay	0	5¼	

No.	LITHOLOGY.	Thickness. ft. in.	ORGANIC REMAINS; AND LOCAL NAMES OF THE BEDS.
13.	Limestone	0 3	"Paving-stone." A few Insects, and *Pholidophorus Stricklandi*, Ag.
14.	Clay	0 10¼	
15.	Limestone	0 3¼	"Bottom rock." More Insects here than in all the other beds collectively.
16.	Clay	0 ·8	
17.	Limestone ....3 in. to	0 6	"Hoggs." *Tetragonolepis angulifer*, Ag. (Warwick Mus.)
18.	Strong hard clay	0 3½	
19.	Argillaceous limestone; imperfect stone	0 3	"Ruskin." No fossils in this quarry.
20.	Laminated clay	1 6	
21.	Fragmentary shelly limestone	0 1½	"Grizzle bed." Saurian bones, Fishes' teeth and scales, *Ammonites planorbis*, *Lima punctata*, *Cardium*, and *Ostrea liassica*; spines of *Cidaris* and other *Echinidæ* abundant.
22.	Stoney shale.		
23.	Hard limestone	0 ·6	"Blue stone" or "Blocks." *Myacites*, and elytra of *Coleoptera*.
24.	Hard clay	1 3	
25.	Limestone	0 3½	"Grave-stone rock." *Ichthyosaurus* and *Otopteris acuminata*, L. & H.
26.	Clay. Thin hard plates of stone lie in this clay	0 11	
27.	Limestone, underlain by clay. (The clay frequently wanting)	0 0½	
28.	Limestone; inconstant..................	0 6	"Gummerals." *Ostrea liassica*.
29.	Clay.		
30.	Hard gray limestone	0 6	"Fire-stone beds." Saurian remains and *Cardium*.
31.	Clay	0 2	*Modiolia minima*, *Myacites*, and *Ostrea liassica*.
32.	Limestone	0 3	In these limestones and clays only *one* small
33.	Clay	0 2	*Ammonites planorbis* has been found.
34.	Limestone	0 3	
35.	Clay	0 3	
36.	Hard dark limestone............ 1 in. to (This is the bottom bed of the quarry.)	0 10	"The Guinea-bed." Saurian bones, *Avicula longicostata*, Stutch., *Monotis decussata* (?), *Lima punctata*, *Myacites*, n. sp., *Ostrea liassica*, and *Hemipedina, Tomesii*, Wright, in numbers; Coral.

AVICULA CONTORTA BEDS.

No.	LITHOLOGY.	Thickness. ft. in.	ORGANIC REMAINS; AND LOCAL NAMES OF THE BEDS.
37.	Thick clay-bed; yellowish blue; breaking in angular fragments	8 0	[Belonging to the zone of *Avicula contorta*.]
38.	Dark ferruginous clay, with conchoidal fracture	8 0	Estheria bed. *Estheria minuta*.
39.	Clay	?	"Clear blue blocks."
40.	Laminated clays.........................	3 0	

No.	LITHOLOGY.	Thickness. ft. in.	ORGANIC REMAINS; AND LOCAL NAMES OF THE BEDS.
41.	Light-coloured sandstone; micaceous	0 1	*Pullastra arenicola*, Strickl.
42.	Brown clay	0 2	
43.	Sandstone; micaceous	0 2	*Pullastra arenicola*, Strickl.
44.	Dark shaly clay	0 6	
45.	Soft sandstone	0 1	
46.	Black clay	0 3	
47.	Ferruginous vein, sandy	?	
48.	Gray Keuper marls.		

The beds from No. 37 to No. 48 were found *in situ* in an escarpment at a short distance from the quarry at Binton. It must be understood that the " Guinea-bed" is the lowest bed seen *in situ* in the pit, and that No. 37 occupies its natural position relatively to that bed, although it is not exposed in the Binton section.

Lithology of the Ammonites planorbis *beds.*—The *Am. planorbis* beds at Brockeridge (p. 58) consist of dark, laminated shales, with interstratified beds of marl and limestone. The shales split into very thin laminæ, between which innumerable shells of *Ammonites planorbis* lie closely compressed ; the white, decomposed, pulverulent matter of the shell forming a strong contrast to the dark shales enclosing them. In Somersetshire the rock consists, at Uphill, of shales which greatly resemble those at Brockeridge ; at Watchet, of dark clays which are more indurated and have preserved better the shell-structure : here *Ammonites planorbis* and *Am. Johnstoni* are found with the iridescent nacreous layer of their shells beautifully preserved. At Street the rock is a light-yellowish clay, with bands of grayish limestone and marl beneath, and in Worcestershire at Stren-sham, and in Warwickshire at Binton, similar lithological conditions prevail.

The White Lias series of the section at Saltford (see p. 64) represents in part the *Am. planorbis* beds : here also the relations of that zone to the Saurian beds below, and to the *Am. Bucklandi* beds above, are well shown. In Dorsetshire the *Am. planorbis* beds are represented by the upper portion of the White Lias so well exposed in the large quarries at Up-Lyme, and in the coast sections at Charlton and Pinhay Bays. The White Lias is raised at Up-Lyme for caustic lime; it consists of a fine-grained, cream-coloured limestone, apparently fit to be used as a lithographic stone. The two principal quarries afford the necessary details. Mr. Webb's quarry shows—

In the uppermost portion, thin bands of gray limestone interstratified with shales ; in these are found *Ammonites planorbis* and *A. Johnstoni ;* in two thick beds of dark, shaly clay are numerous spines and plates of sea-urchins, as *Cidaris Edwardsii*, Wright, *Pseudo-diadema lobatum*, Wright, *Hemipedina Bechii*, Brod., *Hemipedina Bowerbankii*, Wright. These same shales are found at low water-mark at Pinhay Bay, and they have yielded nearly all the Echinidæ said to be found in the Lower Lias at Lyme Regis.

Beneath the lower bed of the Cidaris shales are several thin beds of light-coloured limestone, locally called Whetstones, and separated by intermediate shales; then in descending order come the beds known as Grey Burr, Rotten Burr, Fire-stones, Cliff-ledge, Half-foot, One-foot, Red Size, and Anvil-ledge, all separated by thin bands of shale; on the surface of the limestones, and in the shales are many fossils, among which *Ostrea liassica* forms the dominant shell. The shale above Anvil-ledge contains great numbers of *Pullastra arenicola*, Strickland, apparently indicating a change of condition in the series of beds which lie below this fossil band. The beds from the Whetstones to the Pullastra shales represent the Ostrea series, and from the circumstance of the exposed edges of the rocks having weathered into a cream colour, they form the upper portion of the White Lias. The lower portion of this formation differs both lithologically and palæontologically from the upper portion; it is a compact, earthy limestone, with conchoidal fracture, cream-coloured, and close-grained; many of the beds are so fine that they might be used as lithographic stone. Beneath Anvil-ledge are three feet of light-coloured, rubbly beds, containing *Modiola psilonoti, Ostrea liassica, Myacites musculoides?* resting upon eighteen inches of White Lias; then follow a series of irregular beds, with thin partings, twelve feet in thickness, which overlie a bed, twenty-one inches thick, of fine, white limestone, resting on a like thickness of shale; beneath this is a bed of smooth, regular, fine, white limestone, six feet thick; then a bed of shale; and at the base of the series is a band of Cotham marble or Landscape stone. The lower portion of the White Lias from the Pullastra shales downwards represents, I believe, the upper part of the zone of *Avicula contorta;* there are many fossils in the limestones, which have not yet been determined. I have found casts of *Cardium Rhæticum, Monotis, Pullastra arenicola,* and shells of *Pecten Valoniensis.* Unfortunately the fossils are mostly in the form of moulds, and for this reason we must wait until good specimens are obtained. The lower portion of the White Lias series is only seen in Mr. Fowler's quarry at Up-Lyme.

The coast-section at Pinhay Bay is a complete repetition of the Up-Lyme quarries; the Cidaris shales are here well exposed at low water during spring tides, and from thence are obtained all the *Echinidæ* sold in Lyme Regis. I know of no *Asteriadæ* in these beds.

In Gloucestershire this zone is well exposed at Brockeridge Common, at Wainlode, in a quarry on the right-hand side of the Gloucester Road, between Hartpury and Ashelworth, and at Elmore, in quarries near the Old Kennel.

In Glamorganshire, it is seen in the fine coast-section at Penarth Head. In Somersetshire, in the cutting of the Great Western Railway at Saltford; in the Uphill Cutting on the Bristol and Exeter Railway; in the coast-section at Watchet; and in all the quarries at Street.

In Worcestershire it is found at Strensham; and in Warwickshire at Binton, Grafton, and Wilmcote.

It is likewise found at Robin Hood's Bay, on the coast of Yorkshire; the beds here lie below low-water mark, and large blocks, frequently washed up by the tide, are literally crowded with *Ammonites planorbis*, known at Scarborough and Whitby as *Am. erugatus*, Bean.

The Coral-bed at Lussay, Isle of Skye,[1] probably represents the zone *Am. planorbis*, as I found the same species as the Hebridean coral in the light-coloured clays with *Am. planorbis* at Street.

This lowest Ammonite-zone has, therefore, a wide geographical distribution throughout the Lower Lias of the northern, midland, and southern counties of England, and it retains the same relative position in the Lower Lias of France, Germany, and Switzerland.

Fossils of the Ammonites planorbis *beds.*—The fauna of this zone is very limited; at present I know only the following species :

Ichthyosaurus intermedius, *Conyb.*
— tenuirostris, *Conyb.*
— communis, *Conyb.*
Plesiosaurus Hawkinsii, *Conyb.*
— Etheridgii, *Huxl.*
— rugosus, *Ow.*
— dolichodeirus, *Conyb.*
— megacephalus, *Stutch.*
Dapedius.
Pholidophorus leptocephalus, *Ag.*
— Stricklandi, *Ag.*
Ammonites planorbis, *Sow.*
— Johnstoni, *Sow.*
Lima punctata, *Sow.*

Lima gigantea, *Sow.*
— pectinoides, *Sow.*
Cardinia crassiuscula, *Sow.*
Unicardium cardioides, *Phil.*
Ostrea liassica, *Strick.*
Myacites musculoides? *Schloth.*
Rhynchonella variabilis, *Schloth.*
Cidaris Edwardsii, *Wr.*
Pseudodiadema lobatum, *Wr.*
Hemipedina Bechei, *Brod.*
— Bowerbankii, *Wr.*
— Tomesii, *Wr.*
Isastræa Murchisoni, *Wr.*

2. The Zone of Ammonites Angulatus.

Synonyms.—" Sandige Kalke und Sandsteine mit *Am. angulatus*, Quenst.," ' Flötzgeb.,' 541. " Grès infraliasique (pars)," Dufrénoy and De Beaumont. " Grès liasique, grès de Hettange," Terquem, ' Paléont. du Dép. de la Moselle,' p. 11. " Die Schichten des *Ammonites angulatus*," Oppel, ' Juraformation,' p. 28. " Zone à *Ammonites Moreanus*," Martin, "Pal. Stratigraph. de l'Infra-lias du Départ. de la Côte-d'Or," ' Mém. Géol. Soc. de France,' p. 38, 2nd série, 1860.

The zone of *Ammonites angulatus*, so far as it has been exposed, appears to be imperfectly developed in the British Isles, and from the difficulty experienced in separating its beds from the Bucklandi series, they were grouped with the latter in my memoir. On the Continent, however, and especially in France, this zone forms a very important horizon, and contains a rich fauna. M. Jules Martin,[2] in his valuable memoir on the Infra-Lias of the department of the Côte-d'Or, says, " this zone forms one of the best

---

[1] ' Quart. Journ. Geol. Soc.,' vol. xiv, pp. 4 and 34.

[2] ' Paléontologie Stratigraphique d'Infra-Lias du département de la Côte-d'Or," p. 39, 1860. ' Mém. Soc. Géol. de France,' 2nd série, tom. vii, Mem. No. 1.

characterised palæontological horizons in the Côte-d'Or, and contains the richest and most varied fauna. Limited to two or three yards in thickness, this deposit appears to correspond to a period of animalisation of admirable fecundity. It is from these that we have collected the charming fauna which we last year indicated, and which has such intimate relations with the Hettangian fauna described by M. Terquem." From this zone M. Martin has collected and catalogued—1 *Ichthyosaurus*, 1 *Ichthyodorulite*, 10 sp. of *Cephalopoda*, 63 sp. of *Gasteropoda*, 77 sp. of *Conchifera*, 5 sp. of *Brachiopoda*, 4 sp. of *Echinodermata*, 10 sp. of *Anthozoa*, 3 sp. *Annelida*, and the débris of *Crustacea*. In the department of the Moselle M. Terquem[1] has catalogued 177 species from the same zone. *Ammonites angulatus*, Schloth., is found between Charmouth and Lyme Regis, in dark shale, below the gray concretionary limestone, with a mammillated surface, and likewise in the same strata south-west of the Cob. From these beds I have collected the different varieties of this Ammonite which have been figured by d'Orbigny under the names *Moreanus*, *catenatus*, and *Charmassei*, all of which I regard only as so many different forms of *Ammonites angulatus*, Schloth.

This zone was well exposed in the Harbury cutting of the Great Western Railway, near Warwick, although very few Mollusca besides *Ammonites angulatus* were obtained therefrom; on the spoil banks, even now, some good fragments are occasionally found. On the coast of Yorkshire it occurs near Redcar, where small specimens of this shell, under the local name *Ammonites Redcarensis*, are collected from the clay. In Gloucestershire I have seen small specimens, which were found near Aust and near Gloucester. This zone is likewise exposed in the north of Ireland, in the remarkable Lias district near Portrush.

3. The Zone of Ammonites Bucklandi, or the Lima-Beds.

Synonyms.—"Blue Lias," William Smith, 'Memoir to the Map,' 1815. "Blue Lias Limestone," De la Beche, 'Geol. Trans.,' vol. ii, 2nd series, 1829. "Gryphiten-Kalkstein," Alberti, 'Die Gebirge des König,' Würtemberg, p. 121, 1826. "Liaskalk," Mandelsloh, 'Geol. Profile der schwäbisch,' Alpen, p. 28, 1834. "Calcaire à Gryphée arquée" (pars), Dufrénoy et de Beaumont, 'Mém. Soc. Géol. de France,' 1830. "Grès de Luxembourg (pars. sup.)," Omalius d'Halloy. "Grès de Luxembourg," Dewalque, Descrip. du Lias de la Luxembourg, p. 28, 1857. "Plagiostoma-beds, Lower Lias," Murchison, 'Geol. of Cheltenham,' 2nd ed., p. 49, 1845. "Arietenkalk," Quenstedt, 'Der Jura,' Table, p. 293, 1857. "Die Schichten des Ammonites Bucklandi," Oppel, 'Juraformation,' p. 35, 1856. "Zone of *Ammonites Bucklandi*," Wright, 'Quart. Jour. Geol. Soc.,' vol. xvi, p. 398.

The zone of *Ammonites Bucklandi* (or *Lima*-beds) forms an important subdivision of the Lower Lias. This series attains a great development in the midland counties and in

---

[1] 'Paléontologie du département de la Moselle,' p. 12, 1855.

Glamorganshire, Somerset, and Dorset.   This zone of life is characterised throughout by the prevalence of a number of large *Ammonites* belonging to the natural group *Arietes* (von Buch), and by many *Conchifera* of the genera *Lima* and *Gryphæa*.   In England it everywhere consists of beds of bluish argillaceous limestone, interstratified with beds of marl, shale, and clay of a similar colour.   In some parts of Warwick, Somerset, Dorset, and Glamorgan, this series attains a thickness of 80 feet.

Gloucestershire and Somersetshire.—In Gloucestershire it was partly laid open by the deep cutting of the Dean Forest Railway at Highnam, and it is seen in the Lias limestone-quarries near Tewkesbury, and in the natural escarpments at Frethern and Purton-on-the-Severn.   In Somersetshire it was fully exposed in making the Great Western Railway between Bristol and Bath, and probably at no point were the several beds of the *Lima* series better shown than in the cutting at Saltford, seven miles from Bristol. My friend, Mr. William Sanders, made the following section during the execution of the work which, together with his notes on the fossils contained in the different strata, has been kindly communicated by my friend, Mr. Etheridge.   This section is of great value, inasmuch as the beds are now partially concealed by *débris* and vegetation, and the characteristic fossils can no longer be found in their respective beds.

Section of the Lower Lias Beds at Saltford, seven miles from Bristol, on the Great Western Railway.[1]

No.	LITHOLOGY.	Thickness. feet.	ORGANIC REMAINS.
	Brown gravelly marl	120	
	Beds of laminated shale and clay	110	
	Dark clay, with boulders, and layers of septaria at the top and bottom of the bed, and in the clay between	105	Scales of *Tetragonolepis* and *Belemnites acutus*, Mill.
59.	Gray Lias limestone	100	*Rhynchonella variabilis*, Schloth.
58.	Dark shale		*Belemnites acutus*, Mill.
	Gray Lias limestone.		
	Dark shale		*Ostrea læviuscula*, Sow., *Avicula*, and *Pecten*.
57.	Gray Lias limestone		*Ammonites Conybeari*, Sow., and *Belemn. acutus*, Mill.
	Dark shale.		
56.	Thin limestone-band.		
	Dark limestone		*Nautilus striatus*, Sow., *Am. Conybeari*, Sow., and *Belemnites acutus*, Mill.

[1] This section shows the relative position of the zones of *Ammonites Bucklandi* and *Am. planorbis* and the *Avicula contorta* series in this part of the county, and affords a good type for comparing these zones in Somersetshire with the same groups in other parts of the south of England.

No.	LITHOLOGY.	Thickness. feet.	ORGANIC REMAINS.
55.	Gray Lias limestone .... Dark limestone.	95	*Lima gigantea*, Sow., and *Spirifera Walcotti*, Sow.
54.	Gray Lias limestone. Dark shale.		
53.	Gray Lias limestone. Dark laminated shale.		
52.	Dark-gray Lias limestone		Vertebræ of *Ichthyosaurus*, *Am. Bucklandi*, Sow., and *Spirifera Walcotti*, Sow.
	Dark shale		*Am. Bucklandi*, Sow., *Nautilus striatus*, Sow., and *Spirifera Walcotti*, Sow.
51.	Gray Lias limestone .... Dark shale.	90	
50.	Gray Lias limestone		*Hybodus curtus*, Agass.
	Dark shale		*Pholadomya glabra*, Agass.
49.	Gray Lias limestone		*Nautilus striatus*, Sow. (large), *Am. Brookii*, Sow., and fossil wood.
	Dark shales		*Am. Conybeari*, Sow., and *Am. Bucklandi*, Sow.
48.	Gray Lias limestone		*Pleurotomaria similis*, Sow., and *Lima gigantea*, Sow.
	Dark shales		*Am. Bucklandi*, Sow., and *Pleurotomaria similis*, Sow.
47.	Gray Lias limestone. Dark shales	85	*Ammonites Conybeari*, Sow.
46.	Gray Lias limestone		*Nautilus striatus*, Sow. (large).
	Dark shales		*Pentacrinus tuberculatus*, Mill., (stem) and *Pecten textorius*, Goldf.
45.	Dark-gray limestones.......................		*Ichthyosaurus communis*, Conyb.
	Dark shales		*Gryphæa incurva*, Sow., and *Pentacrinus tuberculatus*, Mill.
44.	Gray limestone. Dark shales.		
43.	Gray limestone		*Ichthyosaurus communis*, Conyb., and *Am. Conybeari*, Sow.
	Dark shales		*Pinna Hartmanni*, Ziet., and *Gryphæa incurva*, Sow.
42.	Bluish limestone	80	
41.	Thirteen or fourteen lime-stone bands, with irregular surfaces; some nodular, with partings of clay and shale ...	75	*Pholadomya glabra*, Ag., and *Lima*, n. sp., with large ribs, *Gryphæa incurva*, Sow., and *Rhynchonella variabilis*, Schl.
40.	From sixteen to eighteen beds, comprising 20 inches of stone	70	*Pholadomya glabra*, Ag., and *Lima gigantea*, Sow.
39.	Fourteen beds of limestone and clay......		*Pecten textorius*, Schl. *Pholadomya glabra*, Ag., and *Pleurotomaria similis*, Sow.
38.	Eight beds of limestone and clay		*Lima pectinoides*, Sow., and *Cardinia ovalis*, Stutch.
37.	Thirteen beds of limestone and clay; the limestones irregular, water-worn, and nodular...............................	65	*Pholadomya glabra*, Ag., *Rhynchonella variabilis*, Schl., and *Ostrea*.
	Dark laminated shales		*Ostrea*.
36.	Gray limestone.		

No.	LITHOLOGY.	Thickness. feet.	ORGANIC REMAINS.
	Dark shales.		
35.	Grayish limestone	60	
34.	Ten beds of shales and limestone ; septaria in the lower beds.		
33.	Thin gray limestone.		
	Thick dark clay.		
32.	Thin band of limestone.		
	Dark clay.		
31.	Thin band of limestone	50	
	Thick dark shales.		
30—25.	Six beds of limestone, alternating with six thicker beds of clay	43	

White Lias Series, 32 feet in thickness.

24.	Light-cloured limestone.		
	Dark-coloured shale.		
23.	Light-coloured limestone.		
	Dark shale.		
22.	Thick White Lias....................	40	
21—12.	Twelve beds of White Lias, separated by thin bands of clay	35	*Pinna Hartmanni*, Ziet., and *Unicardium cardioides*, Phil.
11, 10.	Four beds of limestone, becoming thin and rubbly beneath, and nodular at the base	30	*Pecten textorius*, Goldf., and *Pholadomya glabra*, Ag., *Modiola Hillana*, Sow., and *Avicula* (small).
9.	Cotham marble or Landscape-stone	25	

Avicula contorta Series, 25 feet in thickness.

	Black shales.		
8.	Band of limestone	20	
7.	Nodular limestone.		
	Black shales	15	Fishes' scales ; layers of compressed *Pullastra arenicola*, Strickl.
6.	Dark limestone	10	*Pecten Valoniensis*, Defr., and *Avicula contorta*, Portl.
	Dark shale.		
5.	Dark limestone		*Pullastra arenicola*, Strickl.
	Dark shale.		
4.	Greenish brown soft marl.		
	Marlstone.		
3.	Pale-bluish clay, with plant-like fibres	5	
2.	Buff-coloured clay.		
1.	Gray sandy marlstone, with ferruginous spots.		
	New Red Marl.		

Lyme Regis.—The zone of *Ammonites Bucklandi* is admirably exposed in the coast-section at Lyme Regis, Dorset, both in the Church Cliffs and at Pinhay Bay, where the beds consist of a series of gray limestones, from 2 to 10 inches in thickness, varying from earthy to compact, and alternating with marls and shaly beds—either seams of a few inches, or beds of many feet in thickness. The following section, from the lowest stratum on the shore to Broad Ledge, which may be considered as the uppermost bed of the *Am. Bucklandi* or *Lima* series, affords a correct view of the stratigraphical order of these strata and of the fossils they contain.

Section of the Ammonites Bucklandi *or* Lima *beds from Broad Ledge to the shore at Lyme Regis.*

	No.	LITHOLOGY.	ft.	in.	ORGANIC REMAINS.
Am. Turneri Beds.	1.	Dark-gray limestone. "Broad Ledge" or "Table-bed"	3	6	*Rhynchonella variabilis,* Schloth, in masses.
	2.	Dark marls and shales, with bands of clays	15	0	*Ichthyosaurus communis,* Conyb., *I. platyodon,* Conyb., *Ammonites semicostatus,* Y. & B., and *Rhynchonella variabilis,* Schl.
	3.	Gray limestone	0	4	*Ammonites Turneri,* Sow., and *Am. semicostatus,* Y. & B.
	4.	Dark slaty marls	4	0	
Am. Bucklandi or Lima Beds.	5.	Dark-gray limestone	1	0	*Lima gigantea,* Sow., *L. antiquata,* Sow., and *Rhynchonella variabilis,* Schloth.
	6.	Black shales, with partings of gypsum	2	6	*Ichthyosaurus communis,* Conyb. (in the "fire-stone beds" west of the Cobb).
	7.	Dark-grayish limestone	0	10	*Lima gigantea,* Sow., *L. antiquata,* Sow., and *Rhynchonella variabilis,* Schloth.
	8.	Dark-shale	2	0	*Gryphæa incurva,* Sow.
	9.	Hard gray limestone. "Gray Ledge"	1	3	Fin-spines of *Hybodus, Rhynchonella variabilis,* Schloth, and *Pentacrinus tuberculatus,* Mill.
	10.	Dark shaly marls	2	0	*Ichthyosaurus platyodon,* Conyb.
	11.	Gray limestone	0	6	Spines of *Pseudo-diadema,* and other *Echinidæ.*
	12.	Dark indurated shale.........	3	6	*Ichthyosaurus platyodon,* Conyb.
	13.	Bluish limestone	1	0	*Gryphæa incurva,* Sow., *Rhynchonella variabilis,* Schloth, and *Lima antiquata,* Sow.
	14.	Dark shales, containing indurated imperfect limestone	1	6	*Ichthyosaurus communis,* Conyb., *I. platyodon,* Conyb., *Pentacrinus tuberculatus,* Mill., and *Lima gigantea,* Sow.
	15.	Bluish limestone	0	10	
	16.	Dark indurated clay	1	3	*Gryphæa incurva,* Sow., and fragments of *Pentacrinus tuberculatus,* Mill.
	17.	Gray limestone, with the plant-bed at the top	0	6	*Otopteris obtusa,* L. & H., and *Araucarites peregrinus,* Sternb., in the plant-bed.
	18.	Dark-bluish limestone	1	6	*Ammonites Conybeari,* Sow., and *Rhynchonella variabilis,* Schloth.
	19.	Dark shale.........	0	8	*Gryphæa incurva,* Sow.

	20.	Dark-grayish limestone............	0 10	*Ammonites Bucklandi,* Sow., and *Am. rotiformis,* Sow.
	21.	Dark shale	0 8	
	22.	Gray limestone	0 4	
	23.	Dark shale	1 0	*Ichthyosaurus tenuirostris,* Conyb.
	24.	Dark-grayish limestone	1 4	*Ichthyosaurus communis,* Conyb., skulls and bones of other species, and *Rhynchonella variabilis,*Schloth.
	25.	Hard shale, forming "Quick Ledge".............................	1 6	*Ichthyosaurus communis,* Conyb., and *I. intermedius,* Conyb.
	26.	Blue limestone	0 6	*Ammonites Bucklandi,* Sow., and *Lima gigantea,*Sow.
	27.	Dark shale	0 8	*Gryphæa incurva,* Sow., and *Rhynchonella variabilis,* Schloth.
	28.	Concretionary limestone (surface mammillated)	0 4	
	29.	Dark-gray shale....................	0 8	*Ammonites angulatus,* Schloth.
	30.	Grayish limestone	0 6	*Lima gigantea,* Sow., and *L. antiquata,* Sow.
	31.	Dark indurated shale..............	0 9	*Ammonites angulatus,* Schloth (large specimens). *Gryphæa incurva,* Sow., small and dwarfed.
	32.	Hard gray limestone	0 7	*Lima gigantea,* Sow., and *L. antiquata,* Sow.

(left margin, vertical): Am. Bucklandi or Lima Beds.

The shingle of the shore covers the lower beds.

Coast of Glamorganshire.—In Glamorganshire there is an extensive exposition of the *Am. Bucklandi* and *Lima* series for the distance of sixteen miles along the coast, from Penarth Head, by Barry Island, Aberthaw, and Dunraven Castle to the mouth of the River Ogmore, where the Lower Lias rests on upturned beds of Carboniferous Limestone. The strata chiefly laid bare by the sea are those containing *Lima gigantea* and *Gryphæa incurva.* At Cowbridge the same lithological relations are observed, and the Lower Lias here rests on Carboniferous Limestone.

At Penarth Head, however, the relation of the *Am. Bucklandi* series to the *Am. planorbis* and *Avicula contorta* beds is much better seen than at any other part of the Glamorganshire coast.

Fossils of the zone of Ammonites Bucklandi.—The fossils of the zone of *Ammonites Bucklandi* are numerous, and in general in a good state of preservation.

Ichthyosaurus communis, *Conyb.*
— intermedius, *Conyb.*
— platyodon, *Conyb.*
— tenuirostris, *Conyb.*
Ichthyodorulites of Hybodus.
Ammonites Bucklandi, *Sow.*
— Conybeari, *Sow.*
— rotiformis, *Sow.*
— angulatus, *Schloth.*
— Greenoughii, *Sow.*
— tortilis, *d'Orb.*

Nautilus striatus, *Sow.*
Pleurotomaria similis, *Sow.*
Ostrea irregularis, *Münst.*
Gryphæa incurva, *Sow.*
Unicardium cardioides, *Phil.*
Pecten textorius, *Schloth.*
Lima gigantea, *Sow.*
— antiquata, *Sow.*
— pectinoides, *Sow.*
Modiola Hillana, *Sow.*
Avicula Sinemuriensis, *d'Orb.*

Pinna diluviana (*Zieten*, pl. 55, fig. 6).

Pholadomya glabra, *Agass.*

Terebratula psilonoti, *Quenst.*

Rhynchonella variabilis, *Schloth.*

Spirifera Walcotti, *Sow.*

Pseudo-diadema (spines).

Cidaris Édwardsii, *Wr.*

Pentacrinus tuberculatus, *Mill.*

Isastræa Murchisoni, *Wr.*

3. The Zone of Ammonites Turneri.

Synonyms.—"Hauptpentacrinitenbank des untern Lias," Quenstedt, 'Flözgeb.,' p. 152, 1843. "Lumachelle de *Pentacrinites basaltiformis*," Marcou, 'Jura salinois,' p. 47, 1846. "Die Schichten des *Pentacrinus tuberculatus*" Oppel, 'Juraformation,' p. 44, 1856. "Tuberculatus-bed," Wright, 'Quart. Journ. Geol. Soc.,' vol. xv, p. 25, 1858. "Marne de Strassen," Dewalque et Chapuis, 'Fossiles de Luxembourg,' 1853. "Marne de Strassen," Dewalque, 'Descrip. du Lias de Luxembourg,' 1857. Zone of *Ammonites Turneri*, Wright, 'Quart. Journ. Geol. Soc.,' vol. xvi. p. 403.

This subdivision of the Lower Lias forms a well-marked zone of life. The beds consist of light-coloured argillaceous limestone, of hard grayish limestone, or of deep-blue, shelly, indurated shale, interstratified with beds of dark-coloured clay. Many of the slabs of limestone are covered with shells and portions of the stem and side arms of *Pentacrinus tuberculatus*, Mill. From one of these slabs, collected at Frethern or Purton, in Gloucestershire, Miller's original specimen of this Crinoid was obtained.

Gloucestershire and Warwickshire.—The zone of *Ammonites Turneri* was exposed at Bredon, in the deep cuttings of the Bristol and Birmingham Railway, from whence many of my specimens were obtained. In the Vale of Gloucester portions of these beds are sometimes laid open in making drains, as at Badgeworth and Hardwick; and many fine slabs are occasionally procured from the river-section at Purton. I know of no locality in Gloucestershire, where the entire series is exposed. My friend, Dr. Oppel, referred the Saurian beds of Brockeridge Common to this series, supposing them to be the equivalent of the Saurian beds at Lyme, which, however, appertain to the zone of *Ammonites Turneri;* the description I have already given of the *Am. planorbis* beds and their correlations prove that the beds at Brockeridge Common represent the *Am. planorbis* series. In Warwickshire the *Am. Turneri* beds constitute the base of what is called in that county the "Cardinia-series," in which are included all the strata of the Lower Lias between the *Am. Turneri* and *Am. raricostatus* beds, and which are characterised by different forms of the genus *Cardinia.*

Dorsetshire.—At Lyme Regis the *Ammonites Bucklandi* or *Lima* series is overlain by thick beds of clay and slaty marls containing many Enaliosaurian skeletons, with numerous fishes, in fine preservation; these strata are known to local collectors as the Fish- and

Saurian-beds. The magnificent specimen of *Ichthyosaurus platyodon*, Conyb., now in the British Museum, came from this clay, as is proved by the impressions of *Am. semicostatus*, Y. & B., which are seen on the matrix. This thick clay-bed is underlain by a thin band of grayish-limestone, in which *Am. Turneri* is found. The following section of this zone at Lyme Regis shows the sequence of the *Am. Turneri* beds at that locality.

Section of the Zone of Ammonites Turneri *at Lyme Regis.*

	LITHOLOGY.			ORGANIC REMAINS.
	No.	ft.	in.	
Beds with *Ammonites Turneri.*	1. Thick limestone, "Broad Ledge"	3	6	*Ichthyosaurus platyodon*, Conyb., *I. communis*, Conyb., *Ammonites semicostatus*, Am. Turneri, Sow., and Fishes.
	2. Black shales, with bands of brown clay. "Saurian- and Fish-bed"	18	0	
	3. Grayish, hard, shelly limestone ...	0	4	*Ammonites Turneri*, Sow., and *Am. semicostatus*, Y.&B.
	4. Dark shales, with indurated bands of imperfect limestone	3	0	
	Grayish limestone			*Lima gigantea*, Sow., *L. antiquata*, Sow., and *Rhynchonella variabilis*, Schloth.

Beds with *Am. Bucklandi* and *Lima gigantea*. (See p. 68.)

Fossils of the Zone of Ammonites Turneri.[1]

Ichthyosaurus platyodon, *Conyb.* (British Museum).
— intermedius, *Conyb.* (Warwick Museum).
— communis, *Conyb.* (British Museum).
Ammonites Turneri, *Sow.*
— semicostatus, *Y. & B.*
— Bonnardi, *d'Orb.*
Turbo.
Lima punctata, *Sow.*
— gigantea, *Sow.*
— pectinoides, *Sow.*

Cardinia ovalis, *Stutch.*
Ostrea.
Avicula inæquivalvis, *Sow.*
Pecten textorius, *Schloth.*
— glaber, *Hehl.*
Astarte consobrina, *Dewal.*
Crenatula, nov. sp.
Plicatula spinosa, *Sow.*
Gervillia lanceolata, *Sow.*
Gryphæa incurva, *Sow.*
Cidaris Edwardsii, *Wr.*
Pseudodiadema spines.
Pentacrinus tuberculatus, *Miller.*

[1] I have omitted the fossil Fishes found in the Lias at Lyme Regis, as I was unable to ascertain with sufficient accuracy the beds from which the different species were collected; a large majority of them, however, I believe, came from this zone of life.

4. The Zone of Ammonites obtusus.

Synonyms.—"Marston-Marble," Sowerby, 'Min. Con. Suppl. Index,' vol. i, 1812. "Ammonite-bed (Lower Lias)," Murchison, 'Geol. of Cheltenham,' 2nd edit., p. 42, 1845. "Turneri-Thone," Quenstedt, 'Flözgeb. Württembergs,' p. 540. "Sable d'Aubange (pars infer.)," Dewalque et Chapuis, 'Luxembourg,' p. 12. "Grès de Virton (pars infer.)," Dewalque, 'Lias de Luxembourg,' p. 48. "Die Schichten des *Ammonites obtusus*," Oppel, 'Juraformation,' p. 50. "Indurated marl and limestone-beds," De la Beche, "Section," &c., 'Geol. Trans.,' 2nd ser., vol. ii. Zone of *Ammonites obtusus*, Wright, 'Quart. Journ. Geol. Soc.,' xvi, p. 404.

Gloucestershire and Warwickshire.—The beds constituting this zone are well developed in the Vale of Gloucester, and were exposed in the deep cuttings of the Bristol and Birmingham Railway, especially near Bredon, from whence the best collection of the fossils of these beds was obtained. The rocks consist of dark-gray or bluish shales and clays, with irregular and inconstant beds of dark-gray argillaceous limestone, the shales being in parts nodular or laminated, the clays thick and tenacious; the nodular portions of the shales were very fossiliferous. This zone forms part of the Cardinia-bed of the local geologists of Warwickshire, where it appears to be exposed in several localities. Mr. Tomes' collection contains some very fine specimens of *Ammonites obtusus, Am. multicostatus, Am. Brookii,* and *Am. Sauzeanus,* d'Orb., obtained from the *Am. obtusus* beds; and Mr. Kershaw's cabinet contains a series of *Am. Sauzeanus,* d'Orb., from Darlingstoke, near Shipton-on-Stour. Mr. Etheridge has collected this species at Horfield, near Bristol.

Dorsetshire.—At Lyme Regis the zone of *Ammonites obtusus* attains a considerable thickness, and is well shown in the coast-section. The strata rise on the shore about half a mile west of Charmouth, they consist of thick beds of dark marls, which rest upon the Table-bed, formed by Broad Ledge. The lower part of these marls contains numerous compressed *Ammonites Birchii,* Sow., and layers of nodules forming cement-stones. Above these succeed shales and clays, thin bands of limestone, and thick beds of shale and marls with mudstones. Above these again are inconstant bands of limestone containing septaria, in which gigantic examples of *Am. obtusus, Am. stellaris,* and *Am. Brookii* are found. The following section shows the relative position of these beds.

Section from Broad Ledge to Cornstone Ledge, near Charmouth.

	LITHOLOGY.		ORGANIC REMAINS.
No.		ft. in.	
1.	Dark-gray limestone. "Cornstone Ledge."		
2.	Dark-bluish marls	20 0	
3.	Dark-grayish limestone	0 10	*Ichthyosaurus platyodon*, Conyb., and *I. intermedius*, Conyb. *Ammonites Birchii*, Sow.
4.	Dark clays.		
5.	Dark limestone, with septaria.		*Nautilus striatus*, Sow., *Ammonites Brookii*, Sow., and *Am. stellaris*, Sow. (very large).
6.	Dark shale.		
7.	Dark limestone. "Upper Cement-bed."		
8.	Dark shales, containing mudstone nodules at the base.		*Scelidosaurus Harrisonii*, Owen. *Inoceramus*.
9.	Thin band of limestone. The "Pentacrinite-bed."		*Extracrinus Briareus*, Mill.
10.	Dark shales.		
11.	Dark limestone.	"Fire-ledge."	
12.	Dark shale.		
13.	Dark limestone.		
14.	Dark shale. "Split-ledge."		
15.	Dark limestone		*Ammonites planicosta*, Sow., and *Am. Smithii*, Sow.
16.	Dark shales		Saurian skeletons.
17.	Grayish limestone		*Ammonites obtusus*, Sow., and *Am. Birchii*, Sow., crystallised, forming the "Tortoise-ammonites."
18.	Dark marls, with nodular masses	20 0	The nodules of these lower Cement-beds contain Saurian remains. *Pentacrinus*, n. sp.
19.	Dark indurated shale and limestone. "Broad Ledge."	4 0	This bed overlies the Lima-series east of Lyme-Regis.

The zone of *Ammonites obtusus* probably attains a thickness of from 80 to 100 feet; its actual measurement would be a matter of difficulty, from the manner in which the marls have covered over the bands of limestone; hence the imperfection of our estimate.

In the lower slaty marls are numerous compressed *Ammonites Birchii*, which fall to pieces when removed from the matrix. Higher up (No. 17) the same species is found in fine preservation, with *Ammonites obtusus*. Here the shells are replaced, and the septa filled, with crystallised carbonate of lime. These beautiful specimens are the "Tortoise-ammonites" of local collectors. About 40 or 50 feet above the latter an irregular band of limestone (5) is seen projecting from the cliff, which contains nodules with very large specimens of *Ammonites obtusus*, Sow., *Am. stellaris*, Sow., and *Am. Brookii*, Sow. Most of the nodules have a septarian structure, the veins of spar intersecting and distorting the form of the Ammonites.

Below the Ammonitiferous nodules (5 of the section) other bands of clay and marl (6 to 14) succeed. In one of these (9) are thin layers of Crinoidal limestone, on the surface of which magnificent specimens of *Extracrinus Briareus*, Mill., are found, with their plant-like arms laid out in all directions, and generally coated with sulphuret of iron. The remarkable Liassic Dinosaurian *Scelidosaurus Harrisonii*, Ow., so fully figured and described in the Palæontographical Society's volume for 1859 was discovered some years ago by Mr. Samuel Clarke, of Charmouth, in the dark shales of bed No. 8, above the mudstones.

Fossils of the Zone of Ammonites obtusus.

Scelidosaurus Harrisonii, *Owen.*	Ammonites Smithii, *Sow.*
Ammonites obtusus, *Sow.*	—— Birchii, *Sow.*
—— Brookii, *Sow.*	Nautilus striatus, *Sow.*
—— stellaris, *Sow.*	Belemnites acutus, *Mill.*
—— planicosta, *Sow.*	Pleurotomaria Anglica, *Sow.*
—— Dudressieri, *d'Orb.*	Inoceramus, n. sp.
—— Sauzeanus, *d'Orb.*	Extracrinus Briareus, *Mill.*

5. The Zone of Ammonites oxynotus.

Synonyms.—"Oxynoten-Schichte," Fraas, 'Württemb. naturw. Jahreshefte,' 1847, p. 206. "Oxynotenlager," Quenstedt, 'der Jura,' p. 293, 1858. "Die Schichten des *Ammonites oxynotus*," Oppel, 'die Juraformation,' p. 54, 1856. "Oxynotus-bed," Wright, 'Quart. Journ. Geol. Soc.,' vol. xiv, p. 25, 1858. "Zone of *Ammonites oxynotus*," Wright, 'Quart. Jour. Geol. Soc.,' 1860, vol. xvi, p. 406.

In Gloucestershire this zone consists of beds of dark clays, which often contain much sulphuret or peroxide of iron, all the fossils found in the clay being either highly pyritic or charged with the peroxide of that metal. This bed was exposed in the cuttings of the Bristol and Birmingham, and Great Western Railways, at Lansdown, near Cheltenham, and in excavating the new docks at Gloucester; I have likewise collected its characteristic fossils at other localities in the Vale of Gloucester.

In Dorsetshire a variety of *Ammonites oxynotus*, Quenst., is found in a thin bed of dark, pyritic marl between Charmouth and Lyme Regis, near Black Venn. It is here collected with other species, which properly belong to a higher bed; the falling down of the upper marl, by the decay of the bank, makes it difficult to separate the beds.

At Robin Hood's Bay, on the Yorkshire coast, the relative position of this zone to the beds with *Ammonites obtusus*, Sow., below, and *Ammonites raricostatus*, Ziet., above, are seen in the cliff near the point where the road leads up to the Alum-works. At this locality the *Am. oxynotus* bed is about 20 feet above the clays with *Am. obtusus*.

The form of *Ammonites oxynotus*, Quenst., collected near Cheltenham, exactly resembles the original type of this Ammonite found in Würtemberg. I possess a series of this species, kindly sent me in exchange, from the Royal Museum of Stuttgard, by Professor Fraas which could not be distinguished from ours if they were not previously marked for identification.

Fossils of the Zone of Ammonites oxynotus.

Ammonites oxynotus, *Quenst.*	Plicatula ventricosa, *Münst.*
— bifer, *Quenst.*	Modiola minima, *Sow.*
— lacunatus, *Buck.*	Arca oxynoti, *Wr.*, n. sp.
Nautilus striatus, *Sow.*	Leda Romani, *Oppel.*
Belemnites acutus, *Mill.*	Acrosalenia minuta, *Buck.*
Pleurotomaria Anglica, *Sow.*	Muricated spines of Cidaris.

6. THE ZONE OF AMMONITES RARICOSTATUS.

Synonyms.—" Hippopodium-bed (in part)," Murchison's ' Geology of Cheltenham,' 2nd ed., by Buckman and Strickland, p. 44. " Raricostatenschicht," Fraas, ' Wurttemb. naturw. Jahreshefte,' 1847, pl. 3. " Raricostatenbank," Quenstedt, 1856, ' der Jura,' p. 292. " Die Schichten des *Ammonites raricostatus*," Oppel, 1856, ' die Juraforma-tion,' p. 56. " Raricostatus-bed," Wright, ' Quart. Jour. Geol. Soc.,' vol. xiv, p. 25. " Zone of *Ammonites raricostatus*," Wright, ' Quart. Jour. Geol. Soc.,' vol. xvi, p. 407.

The beds forming this zone are exposed in several brick-fields in the vicinity of Cheltenham. They consist of dark-coloured clays, more or less impregnated with the peroxide of iron. In an excavation recently made near Marle Hill, for the purpose of obtaining clay to make bricks for the town-sewers, the following section was obtained. The beds are enumerated in descending order.

No.		ft.	in.
1.	*Gryphæa-bed;* a hard, ferruginous clay, which broke into fragments, and contained a great many specimens of *Gryphæa obliquata*, Sow. .... 3 ft. to	4	0
2.	*Coral-band;* a thin seam of lightish-coloured, unctuous clay, containing a great many small, sessile Corals, *Thecocyathus rugosus*, Wr., most of which appeared to have been attached to the curved valves of the *Gryphææ* .... 1 in. to	0	1½
3.	*Hippopodium-bed;* a stiff, dark-coloured clay, in some parts ferruginous; containing *Cardinia Listeri*, Sow., and *Hippopodium ponderosum*, Sow., in consider-able numbers... from 8 ft. to	10	0
4.	*Ammonite-bed;* a dark, ferruginous clay, containing selenite, with the peroxide and sulphuret of iron, and great numbers of a highly pyritic brood of *Ammonites*, likewise *Am. raricostatus, Am. armatus*, and the other species of the list .... } Not ascertained.		

In the parish of Cleeve, near Cheltenham, the same beds were formerly worked for brick-earth; the finest specimens I have collected of *Cardinia Listeri*, Sow., *Hippopodium ponderosum*, Sow., *Ammonites raricostatus*, Ziet., and *Pleurotomaria Anglica*, Sow., were obtained therefrom. In the railway-cutting at Bredon the same beds were likewise laid open, and yielded a rich series of the characteristic fossils. In Warwickshire the railway-cutting at Honeybourne exposed the same beds; and here also the Coralband contained a considerable number of *Thecocyathus rugosus*, Wr.

At Lyme Regis, in Dorsetshire, this zone is found near Black Venn. Some of the beds contain a considerable quantitity of pyrites, so much so that during the winter months they are worked for that mineral, when their characteristic *Ammonites* are collected in considerable numbers; unfortunately these fossils are so much charged with pyrites that they are with difficulty preserved.

At Robin Hood's Bay, on the coast of Yorkshire, this zone is seen resting on the underlying clays with *Ammonites oxynotus*, and overlain by thick clays containing *Ammonites Jamesoni*, Sow. In all these localities there appears to be an absence of limestone-layers; clay, more or less impregnated with iron, constituting the entire beds.

Fossils of the Zone of Ammonites raricostatus.

Belemnites acutus, *Mill.*	Ostrea, raricostata, *Wr.*
Nautilus striatus, *Sow.*	Gryphæa obliquata, *Sow.*
Ammonites raricostatus, *Ziet.*	Cardinia Listeri, *Sow.*
—— armatus, *Sow.*	Hippopodium ponderosum, *Sow.*
—— armatus densinodus, *Quenst.*	Anomya pellucida, *Terq.*
—— nodulosus, *Buck.*	Unicardium cardioides, *Phil.*
—— Guibalianus, *d'Orb.*	Pleuromya oblonga, *Wr.*, n. sp.
—— muticus, *d'Orb.* (?)	Rhynchonella variabilis, *Schloth.*
Pleurotomaria similis, *Sow.*	Terebratula numismalis, *Lamk.*
Trochus imbricatus, *Sow.*	Pentacrinus scalaris, *Goldf.*
Chemnitzia parva, *Wr.*, n. sp.	Thecocyathus rugosus, *Wr.*

THE MIDDLE LIAS.

The Middle Lias is well developed in England, and fully exposed in the grand natural sections of the Yorkshire and Dorsetshire coasts. In the Midland Counties it is only partially shown. I therefore select a section of the cliffs east of Charmouth, Dorset, which I made for this work last summer, with the assistance of my friend Mr. Day, as the one that affords the best general view of the whole; the Middle Lias here attains a thickness of about 450 feet, and is divisible into five stages, each characterised by special specific forms; these in ascending order are—1st. The zone of *Ammonites Jamesoni*. 2nd. The zone of *Ammonites Ibex*. 3rd. The zone of *Ammonites capricornus*. 4th. The zone of *Ammonites margaritatus*. 5th. The zone of *Ammonites spinatus*.

Section of Down Cliffs, at Toad's Cove, near Bridport Harbour.

	LITHOLOGY.	Thickness. feet.	ORGANIC REMAINS.
Elevation.			
400 ft.	ZONE OF AMMONITES JURENSIS.		
No. 1.	Brown sands, sometimes micaceous, with large sandstone nodules in layers	70	*Ammonites opalinus*, Rein., at Burton Cliff.
	ZONE OF AMMONITES COMMUNIS.		
2.	Dark-grayish sandy marl, very micaceous	72	Fossils rare, and indeterminable.
300 ft.			
3.	Brownish marly limestone, containing great numbers of *Ammonites serpentinus* and other Upper Lias shells. The Middle Lias comes up to the lower part of this band of stone, for *Ammonitus spinatus* has been found in it by Mr. Day	2½	*Ammonites serpentinus*, Schlot., *A. communis*, Sow., *A. bifrons*, Brug., *Raquinianus*, d'Orb., *Pleuromya unioides*, Roem., *Venus pumila*, Münst., *Rhynch. acuta*, Sow., and *Rhynch. Moorei*, Dav.
	ZONE OF AMMONITES SPINATUS.		
4.	Dark-gray, sandy, micaceous marl	18	*Ammonites spinatus*, Brug. *Belemnites breviformis*, Ziet.
5.	Indurated sand, forming large sandstone blocks	8	
6.	Light-brown sands, more or less indurated, and very micaceous	56	Found no fossils to enable us to determine whether the bed belongs to this, or the lower zone.
200 ft.	ZONE OF AMMONITES MARGARITATUS.		
7.	Bluish marl, which forms a well-defined band in the section	6—8	
8.	Grayish, sandy, laminated marls, with irregular layers of nodules	20	*Ammonites margaritatus*, Mont., *Am. fimbriatus*, Sow., *Belemnites elongatus*, Mill., *Pleurotomaria Anglica*, Sow., *P. expansa*, Sow., *Pleuromya unioides*, Roem., *Pecten æquivalvis*, Sow., *Limea acuticostata*, Münst., *Lima Hermanni*, Münst., *Pinna Hartmanni*, Münst., *Rhynchonella acuta*, Sow., *R. tetrahædra*, Sow., *R. spinosa*, ? Schloth., *Gryphæa gigantea*, Sow.
9.	Foxy-coloured sandstone, with from 12—16 irregular bands of stone forming the "rough bed" of the workmen	40	
10.	Band of Crinoidal limestone		*Pentacrinus subangularis*, Mill.
11.	Gray sandy clay, in parts micaceous	20	
100 ft.	Band of ferruginous septariæ.		
12.	Gray laminated sandy clay	17	

Elevation.	LITHOLOGY.	Thickness. feet.	ORGANIC REMAINS.
No.			
13.	The "Starfish Bed," hard, gray, micaceous sandstone, large blocks from this bed lie on the shore..............	6	*Ophioderma Egertoni*, Brod., *Ammonites fimbriatus*, Sow., *A. margaritatus*, Mont., *Belemnites elongatus*, Mill.
	ZONE OF AMMONITES CAPRICORNUS.		
14.	Gray marls, breaking up into cuboidal masses; in the upper part are several rows of small, fossiliferous nodules; this bed is much thicker, and better seen at Golden Cap.....................	76	*Ammonites capricornus*, Schloth., *A. Henleyi*, Sow., *A. Bechei*, Sow., *A. Davæi*, Sow.
	Base of Down Cliffs		

The gray or micaceous marls attain a great thickness at Golden Cap, where they rest on the Belemnite bed. These marls contain several stages of life, which have not been worked out with sufficient accuracy, to enable me to define the limits of the different zones. Fragments of *Ammonites Jamesoni* have been collected in the lower part of this deposit, and *Ammonites Davæi, capricornus,* and *Bechei,* in the upper. In Gloucestershire beneath the zone of *Ammonites capricornus* two other zones are found characterised by *Ammonites Ibex* and *Ammonites Jamesoni,* and these are likewise doubtless comprised in the gray, micaceous marls at Golden Cap, which here attain so great a thickness.

7. THE ZONE OF AMMONITES JAMESONI.

Synonyms.—" Micaceous marl, in part," De la Beche's section. " Numismalismergel oder Belemnitenmergel," Quenstedt, ' das Flötzgebirge Würtemberg." " Lias Gamma, pars," Quenst., ' der Jura Uebersichtstafel,' p. 293. "Die Schichten des *Ammonites Jamesoni,*" Oppel, ' die Juraformation,' p. 118. " Jamesoni-bed," Wright, ' Quart. Jour. Geol. Soc.,' vol. xiv, p. 25.

In Gloucestershire the beds representing this zone are found only in some deep brick-pits near Leckhampton, in the environs of Cheltenham, from whence I obtained fragments of a large *Ammonites Jamesoni,* Sow., and many of the young forms of this species known as *A. Bronni,* Röm., with *Rhynchonella rimosa,* von Buch.

The same Ammonites are found at Robin Hood's Bay, on the coast of Yorkshire, where these beds at the west of the bay, according to Dr. Oppel, attain a thickness of 100 feet; with *Ammonites Jamesoni* were associated *Ammonites Taylori,* Sow., *Bel. elongatus,* Mill., *Gryphæa obliquata,* Sow., *Pholadomya decorata,* Ag., and *Pinna folium,* Phil.

This zone is well developed in the Island of Pabba, near Skye, in the Hebrides, where *Ammonites Jamesoni* is moderately abundant, if I may form an opinion from the number of specimens of that Ammonite collected at Pabba by Mr. Geikie[1] in a brown, micaceous sand-stone, and which have passed through my hands for determination.

The original type of this Ammonite was collected by Sir R. Murchison from beds of the same age in the adjacent Isle of Mull.

In Dorsetshire this Ammonite is occasionally found; I have not seen a specimen, nor have I been able to determine its bed.

8. The Zone of Ammonites Ibex.

Synonyms.—" Ochraceous Lias," Murchison, ' Geol. of Cheltenham,' 2nd edit., 42. " Die Schichten des *Ammonites Ibex*," Oppel, ' die Juraformation,' p. 122. " Upper marls, pars," De la Beche, ' Geol. of Lyme Regis.' " Ibex-bed," Wright, ' Quart. Jour. Geol. Soc.,' vol. xiv, p. 25.

In the neighbourhood of Cheltenham this zone is often exposed, and from hard, calcareous nodules, imbedded in a light-coloured clay, its characteristic Ammonites are collected; these are *Ammonites Ibex*, Quenst., *A. Maugenesti*, d'Orb., and *A. Henleyi*, Sow., *A. bipunctatus*, Röm., with several Conchifera, as *Crenatula ventricosa*, Sow., *Mytilus scalprum*, Sow., *Arca elongata*, Quenst., *Pinna folium*, Y. and B., *Cardinia attenuata*, Buck., and many other undescribed forms.

In Northamptonshire fine specimens of *A. Ibex*, Quenst., and *A. bipunctatus*, Röm., have been collected near Watford.

9. The Zone of Ammonites-capricornus.

Synonyms.—" Lias Gamma (pars Davæikalk)," Quenstedt, ' der Jura Uebersichstafel,' p. 293. " Schiste d'Ethe," Dewalque, ' Lias de Luxembourg,' p. 55. " Die Schichten des *Ammonites Davæi*," Oppel, ' Juraformation,' p. 126. " Davæi-bed," Wright, ' Quart. Jour. Geol. Soc.,' vol. xiv, p. 25.

This zone is very well developed in England; and wherever the Middle Lias is complete it is found beneath the marlstone, consisting of laminated clays, containing micaceous particles, or the argile is richly charged with peroxide of iron, whilst in other localities it is represented by a brown, micaceous sandstone. The irregular, stony bands found in the clay usually contain a large assemblage of fossils. This zone is likewise especially rich in the remains of Echinodermata, and claims our special interest on that account.

---

[1] ' Quart. Journ. of the Geol. Soc.,' vol. xiv, p. 28.

At Mickelton[1] tunnel the shale which lies below the marlstone was exposed to a considerable extent, and yielded a great number of fossils in a high state of preservation. The shales were in parts arenaceous, and formed thin slabs of a fine, bluish sandstone, on which many of the Echinodermata were found.   There were also large slabs of ironstone, sixteen inches in thickness, almost entirely composed of shells ; the upper and under surfaces in many of them were crowded with fragments of *Pentacrinites* and skeletons of *Asteriadæ*. *Uraster Gaveyi*, Forb., figured in our plate, was discovered on the upper surface of a slab of sandstone, twelve inches in thickness, associated with plates and fragments of *Pentacrinus robustus*, Wr., and several of the Conchifera of the subjoined list, together with compressed shells of *Ammonites capricornus*, Schloth.

All the specimens of *Tropidaster pectinatus*, Forb., *Ophioderma Gaveyi*, Wr., and *Cidaris Edwardsii*, Wr., were found attached to the under side of a thick slab of ironstone, about twenty feet below the surface.   Almost all the specimens show their ventral surface, and most of them have their spines fixed on the spiniferous tubercles to which they belonged ; several of the star-fishes are as well preserved for anatomical description as if they had been prepared from recent specimens for that purpose.

Beds of laminated shale and ferruginous clay, the equivalent in age of those at Mickelton, were cut through in making the deep excavations on Hewlett's Hill for the formation of the reservoirs of the Cheltenham Waterworks Company.   In the laminated clays the shells of the Mollusca and the tests of the Echinodermata were well preserved, but the fossils in the ferruginous clay, although very abundant, soon perished from the large per-centage of iron which the argile contained.

The same bands of rock were likewise laid open at Witcombe Park, near Birdlip.   In constructing the reservoir for receiving the water flowing from the Witcombe Spring for the supply of the City of Gloucester, the fossils were abundant, but mostly fragmentary. In many the shell was preserved, and some fine specimens of *Nautilus striatus* were obtained.

Fossils from the Zone of Ammonites capricornus, *in Gloucestershire.*

CEPHALOPODA.

Belemnites umbilicatus, *d'Blanv.*	Ammonites Henleyi, *Sow.*
— elongatus, *Mill.*	— capricornus, *Schloth.*
— paxillosus, *Schloth.*	— fimbriatus, *Sow.*
Nautilus striatus, *Sow.*	— Davæi, *Sow.*

GASTEROPODA.

Chemnitzia capricorni, *Wr.,* n. sp.	Pleurotomaria Anglica, *Sow.*
Cylindrites capricorni, *Wr.,* n. sp.	— expansa, *Sow.*
Trochus imbricatus, *Sow.*	— undosus, *Schübl.*

---

[1] "On the Railway Cuttings at the Mickleton Tunnel, &c.," by G. E. Gavey, Esq., with sections; 'Quart. Journ. Geol. Soc.,' vol. ix, p. 29.

CONCHIFERA.

Pholadomya ambigua, *Sow.*
— decorata, *Hartm.*
Pleuromya unioides, *Röm.*
Leda rostralis, *Lamk.*
— complanata, *Röm.*
— acuminata, *Goldf.*
— cordata, *Goldf.*
— inflexa, *Röm.*
Astarte capricorni, *Wr.*
Mytilus hippocampus, *Young* and *Bird.*
Cypricardia cucullata, *Goldf.*
Cardinia attenuata, *Stutch.*
Goniomya capricorni, *Wr.*
Cardium truncatum, *Phil.*
Unicardium Janthe, *d'Orbig.*
Cucullæa Münsteri, *Ziet.*

Arca elongata, *Quenst.*
— truncata, *Buck.*
Modiola scalprum, *Sow.*
Limea acuticosta, *Gold.*
Avicula longiaxis, *Buck.*
Monotis inæquivalvis, *Sow.*
Inoceramus ventricosus, *Sow.*
— substriatus, *Goldf.*
Pecten æquivalvis, *Sow.*
— priscus, *Schloth.*
— diversus, *Buck.*
— liasinus, *Nyst.*
Gervillia lævis, *Buck.*
Plicatula spinosa, *Sow.*
Gryphæa cymbium, *Lam.*
Ostrea.

BRACHIOPODA.

Terebratula punctata, *Sow.*
Spirifer rostratus, *Schloth.*
Rhynchonella rimosa, *von Buch.*

Rynchonella variabilis, *Schloth.*
Orbicula scaliforme, *Wr.*, n. sp.
Lingula Beanii, *Phil.*

ECHINODERMATA.

Cidaris Edwardsi, *Wr.*
Acrosalenia, sp.
Pedina, sp.
Uraster Gaveyi, *Forb.*
Tropidaster pectinatus, *Forb.*

Ophioderma Gaveyi, *Wr.*
Ophioderma Brodiei, *Wr.* n. sp.
Pentacrinus robustus, *Wr.*
— punctiferus, *Quenst.*
— subangularis, *Mill.*

On the coast of Yorkshire, as at Staithes, Boulby, and Skinningrave Bay, where the Marlstone series is admirably exposed, the shales with *Ammonites maculatus* (*capricornus*) form the base of the cliffs. "This Ammonite," observes Mr. Hunton,[1] "is constantly found at the junction of the Marlstone with the Lower Lias (zone of *A. capricornus*), which here pass so gradually into each other that it is impossible to determine where the sandstones end and the blue shale begins. I have long sought for *A. maculatus*, Y. & B., (*capricornus*, Schloth.) in the upper and central portions of the Marlstone, but have never found it many feet above the junction beds; and though this and other Ammonites from unequal geographical distribution may be more abundant in one place than in another (*A. maculatus* is in greatest number at Staithes, *A. Hawskerensis* at Hawskerbottoms), yet

[1] 'Trans. Geol. Soc.,' second series, vol. v, p. 218.

11

they constantly maintain an unvariable relative position." It was from the upper shales of this zone at Robin Hood's Bay that *Luidia Murchisonæ*, Will., was obtained, and from beds of the same age at Skinningrave that *Plumaster ophiuroides*, Wr., was extracted. It would appear, therefore, that the northern limit of this stage in England is characterised by species of ASTERIADÆ entirely distinct from those found in the south, although the species of the molluscan fauna are the same throughout.

On the Dorsetshire coast, as near Charmouth, this zone forms the upper portion of the gray micaceous marls, or "the green Ammonite-bed" of local collectors. The Charmouth form of *A. capricornus*, Schl., was figured by Sowerby in his 'Mineral Conchology,' vol. vi, pl. 556, fig. 2, as *A. latæcosta;* and to increase the existing confusion regarding the identity of this species, a mistake was made in the text in reference to the figures—*A. latæcosta* was described as *A. brevispina*, and *vice versâ;* this fact I have verified by an examination of the original specimens. *A. latæcosta*, Sow., is therefore only a variety of *A. capricornus* from the upper portion of the gray marls, whilst *A. brevispina*, Sow., is a distinct species, which I have hitherto only detected among fossils from the zone of *Ammonites Jamesoni* at Pabba. In my description of *Ammonites brevispina*, Sow.,[1] it is stated—"Sowerby's text and the numbers of pl. 556 do not agree; this mistake will mislead the reader, unless he is acquainted with the two Ammonites figured in that plate, for fig. 1 represents *Amm. brevispina*, Sow., and fig. 2, *Amm. latæcosta*, Sow.; the former is a Pabba, the latter a Charmouth, fossil.

Ammonites Loscombi, Sow., is a very abundant form, and which has a very wide range throughout the gray micaceous marls.

11. THE ZONE OF AMMONITES MARGARITATUS.

Synonyms.—"Marlstone and ironstone series," Phillips, 'Geol. of York.,' p. 192. "The Marlstone," Murchison, 'Geol. of Cheltenham,' 2nd edit., p. 37. "Marlstone," Williamson, "Fossil Remains on the Yorkshire Coast," 'Geol. Trans.,' 2nd ser., vol. v, p. 224. "Marlstone," Hull, 'Memoirs of the Geol. Survey,' descrip. of sheet 44, p. 18. "Lias Delta, Amaltheenthon," Quenstedt, 'Flözgeb.,' p. 540. "Marnes à *Ammonites amaltheus* ou *margaritatus*," Marcou, 'Jura Salinois,' p. 50. "Macigno d'Aubange," Dumont, Dewalque et Chapuis, 'Luxembourg,' p. 273. "Die oberen Schichten des *Ammonites margaritatus*," Oppel, 'Juraformation,' p. 133. "Margaritatus-bed," Wright, 'Quart. Journ. Geol. Soc.,' vol. xv, p. 25.

The Marlstone forms a well-known division of the Lias formation. In Gloucestershire it consists, according to Mr. Hull,[2] of two parts—the lower, a series of yellow, gray, and brown sands, with thin bands of calcareous limestone and ferruginous nodules; the

---

[1] 'Quart. Journ. Geol. Soc.,' vol. xiv, p. 28.

[2] 'Memoirs of the Geol. Survey. The Country around Cheltenham,' p. 18.

upper, a rock-bed of an impure limestone, weathering blue or brown in the interior. In the eastern part of the district it is highly ferruginous, and varies from one to ten feet in thickness. The rock-bed forms the surface of the tabulated promontories which produce such picturesque features along the flanks of the Cotteswold Hills, and around many of the outliers, while the steep, fertile banks which descend from the edges of the platforms to the Lower Lias plain are composed of the underlying beds of sand.

At Leckhampton Hill the thickness of this formation is 115 feet, and this is its estimated general thickness around the Cotteswold Hills. The zone of *Ammonites spinatus* is so closely incorporated with the Marlstone rock-bed that it must be included in this stage, of which it forms the uppermost portion.

Fossils from the Marlstone.

VERTEBRATA.

Vertebræ of Ichthyosaurus.

Teeth and scales of fishes.

CEPHALOPODA.

Belemnites compressus, *Stahl.*
— paxillosus, *Schloth.*
Nautilus striatus, *Sow.*
— intermedius, *Sow.*

Ammonites margaritatus, *Montf.*
— Englehardti, *d'Orbig.*
— Normanianus, *d'Orbig.*
— heterophyllus amalthei, *Quenst.*

GASTEROPODA.

Pleurotomaria Anglica, *Sow.*
— undosus, *Schübl.*
— expansa, *Sow.*

Chemnitzia undulata, *Ziet.*
Turbo orion, *d'Orbig.*
Trochus imbricatus, *Sow.*

CONCHIFERA.

Pholadomya ambigua, *Sow.*
Pleuromya unioides, *Röm.*
— rotundata, *Goldf.*
— Alduini, *Brong.*
— donaciforme, *Goldf.*
Ceromya lineata, *Will.*
Arcomya elongata, *Röm.*
Unicardium cardioides, *Phil.*
Cardium truncatum, *Phil.*
Cardinia crassiuscula, *Sow.*
— crassissima, *Sow.*
Goniomya capricorni, *Wr.*

Cypricardia cucullata, *Goldf.*
Modiola scalprum, *Sow.*
Limea acuticosta, *Goldf.*
Monotis inæquivalvis, *Sow.*
Lima Hermanni, *Voltz.*
— duplicata, *Sow.*
— pectinoides, *Sow.*
Pecten diversus, *Buck.*
— æquivalvis, *Sow.*
— cinctus, *Sow.*
Gryphæa gigantea, *Sow.*
Ostrea.

BRACHIOPODA.

Terebratula punctata *Sow.*	Rhynchonella tetrahedra, *Sow.*
— resupinata, *Sow.*	— variabilis, *Schloth.*
— cornuta, *Sow.*	— acuta, *Sow.*
— Edwardsii, *David.*	Lingula Beanii, *Phil.*

ECHINODERMATA.

Hemipedina Jardinii, *Wr.*	Pentacrinus subangularis, *Mill.*

The Marlstone attains a great development in Yorkshire, and is fully exposed in the coast section near Staithes. "The sandy, conchiferous marlstone beds," says Prof. Phillips,[1] "which in Colborn Nab cover the Lower Lias shale, are seen rising with it and contributing to swell the altitude of Boulby and Rockcliff. The lower part of this series is generally the most solid, and projects in broad, compact floors above the Lias. On the surfaces of such beds lie innumerable multitudes of Oysters, Dentalia, Pectens, *Cardium truncatum, Avicula inæquivalvis,* and, more rarely about Staithes, beautiful fossil star-fishes of the genus *Ophiura.*" In Boulby Cliffs[2] the ironstone and Marlstone series consists of—

a. The ironstone bands, which are numerous layers of firmly connected nodules of ironstone, often septiarate, and enclosing dicotyledonous wood, Pectines, Aviculæ, Terebratulæ, and from twenty to forty feet thick.

b. The Marlstone series, consisting of alternations of sandy Lias shale and sandstones, which are frequently calcareous, and generally full of shells. The lower beds are usually most solid, and project from the cliffs in broad floors, covered with Pectens, Cardia, Dentalia, Aviculæ, Gryphææ, &c. The thickness variable from forty to 120 feet.

The molluscan fauna of these beds closely resembles the list already given from the Marlstone of Gloucestershire. Among the Echinodermata, however, are found species which appear to be limited to the Yorkshire series, as *Uraster carinatus,* Wr., *Astropecten Hastingsiæ,* Forb., *Ophioderma Milleri,* Phil., *Aspidura loricata,* Will., *and Ophiura Murravii,* Forb., all collected from the Marlstone series near Staithes.

The zone of *Ammonites margaritatus* is exposed at Toad's Cove, Down Cliffs, near Bridport Harbour, Dorset; its position is shown in the section, p. 77, where the rock-bed contains *Ammonites margaritatus* and many of the same species of Conchifera found associated with that Ammonite, in other Marlstone districts. The star-fish-bed, with *Ophioderma Egertoni,* Brod., and *Ammonites fimbriatus,* Sow., is seen beneath the rock-bed of the Marlstone series.

---

[1] 'Geology of Yorkshire,' p. 101.
[2] Ibid., p. 102.

12. The Zone of Ammonites spinatus.

Synonyms.—Upper portion of the "Marlstone" of English authors. "Marnes à Plicatules," Marcou, 'Jura Salinois,' p. 51. " Amaltheenthone (pars. sup.)," Quenst. " Die Schichten des *Ammonites spinatus,*" Oppel, ' Juraformation,' p. 138. " Spinatus-bed," Wright, ' Quart. Journ. Geol. Soc.,' vol. xiv, p. 25.

This zone is so closely united with the Marlstone that it appears to form its upper portion, lithologically it is a light-coloured friable bed, containing many large fossiliferous nodules. *Ammonites spinatus,* Brug., *Belemnites breviformis,* Ziet., *Lima Hermanni,* Voltz., *Terebratula punctata,* Sow., and *Spirifer rostratus,* Schloth., are the prevailing forms.—

In Down Cliff the spinatus bed lies immediately in contact with the indurated marl of the Upper Lias containing *Ammonites serpentinus,* Reinecke, and other Upper Lias forms (see p. 77).

THE UPPER LIAS.

I include all the marly, argillaceous, and arenaceous deposits found between the Marlstone and Inferior Oolite in the Upper Lias, and group these beds into two stages, each of which contains a special fauna.

The lower zone is, in general, an argillaceous formation, with occasional and inconstant bands of calcareous nodules; the fossils of this division are nearly all specifically distinct from those of the Marlstone on which it rests. The Ammonites of the group CAPRICORNI are all absent from these beds; and in their stead have appeared great numbers of the groups FALCIFERI and PLANULATI. In England one of the most prevailing species of the latter is the *Ammonites communis,* Sow., from which I have derived the name of the stage.

The upper zone in England is essentially an arenaceous formation, and although it possesses some species in common with the zones of *Ammonites communis* below, and *Ammonites Murchisonæ* of the Inferior Oolite above, it nevertheless contains a fauna sufficiently numerous in special forms to justify its separation from the *Am. communis* stage. Most of the Ammonites of the upper zone belong to the group FALCIFERI; a few, however, are common to both, those of the group LINEATI found herein are special to this zone, one of the most characteristic of which is the *Ammonites Jurensis,* Ziet.

13. The Zone of Ammonites communis.

Synonyms.—" Alum shale," Young and Bird, ' Geol. of York.,' p. 133.   " Upper Lias"
of English authors.   " Posidonien-Schiefer," Römer, ' Oolit. Geb.,' p. 5, 1836.   " Lias-
Schiefer," von Buch, ' Jura Deutsch., Berl. Akadem,' 1837.   " Posidonien-Schiefer,"
Quenst., ' Flötzgebirge,' p. 538.   " 9ᵉ Étage, Toarcien (pars infer.)," d'Orbigny, ' Cours.
élément. de Paléontol.,' p. 463.   " Die Schichten der *Posidonomya Bronni*," Oppel,
' Juraformation,' p. 197.   " Communis-bed," Wright, ' Quart. Journ. Geol. Soc.,' vol. xiv,
p. 25.

In Gloucestershire this zone consists of bluish clay containing occasional and irregular
bands of nodular argillaceous limestone resembling " cement stones."   In the escarpments
of the Cotteswold Hills it attains in some places a thickness of from 100 to 200 feet, and
is there interposed between the sands or rock-beds of the Inferior Oolite and Marlstone.   The Upper Lias clay is generally concealed by débris from superior strata, and
its position is therefore most readily ascertained by surface indications, such as springs
and marshes.   As this clay-bed forms a retentive stratum at the base of the superincum-
bent porous strata, the rain, which falls upon the table-land of the Cotteswold Hills, after
saturating the Oolitic rocks and subjacent sands, bursts forth as springs along their slopes
and escarpments, at the junction of these beds with the impervious clay.   All the springs
in this district arising from the drainage of the Inferior Oolite have their origin in this
arrangement of the strata.

On the summits of Bredon, Alderton, Gretton, and Churchdown Hills, all outliers of
the Cotteswolds, we find sections of the lower or basement beds of the Upper Lias ; these
consist in general of the following sub-divisions :

1st. Brown marly clays of variable thickness, according to the extent of denudation of
the upper beds ; they contain many of the fossils of our list.

2nd. A band of nodular argillaceous limestone from six to eight inches in thickness,
called the " Fish bed," this stratum has yielded many interesting remains.   I obtained
from a nodule at Gretton, a large and nearly perfect specimen of *Pachycormus latirostris*,
Ag. ? and from nodules at Alderton, Dumbleton, and Gretton, *Leptolepis concentricus*,
Egert., *Tetragonolepis discus*, Egert., have been extracted.   Wings and elytra of insects
have likewise been found in nodules at Dumbleton and Gretton, of which the most
remarkable is a fine Neuropterous wing belonging to *Libellula Brodiei*, Buck.

3rd. Is a thick bed of bluish mottled clay, several feet in thickness, and more or less
laminated, at Alderton, where I have seen it once well exposed ; it contained a great
many small Gasteropoda, among them were *Cerithium*, *Rostellaria*, *Trochus*, and *Natica ;*
of Conchifera, I found *Arca*, *Leda*, and *Posidonomya ;* of Echinodermata I observed
Acrosalenia crinifera, Quenst., *Pseudodiadema Moorei*, Wr., *Ophioderma*, n. sp., and

fragments of *Pentacrinus*. The shells were compressed moulds, which looked beautiful when the clay was first split open, but as it dried, the fossils unfortunately broke into fragments.

4th. The Leptæna bed is composed of a brown friable marl, one to two inches thick; it contains many species of small Brachiopoda, belonging to the genera *Leptæna*, *Spirifera*, and *Terebratula*, and is separated from the upper beds of the Marlstone by—

5th. A thin band of blue and yellow clay, containing many *Ammonites falcifer*, Sow., *Belemnites acuarius*, Schloth., and Brachiopoda, as *Rhynchonella pygmæa*, Moore. This bed rests upon light-coloured marls of the Spinatus stage.

From a railway cutting near Stroud, I have obtained many fine specimens of *Ammonites serpentinus*, Rein., and during the execution of works for drainage, and the formation of a new road near Nailsworth, a good section of the Upper Lias was exposed; from the bands of limestone numerous fossils were obtained, as *Ammonites communis*, Sow., *A. bifrons*, Brug., *A. falcifer*, Sow., *A. heterophyllus*, Sow., *A. cornucopia*, Y. and B., *A. Lythensis*, Y. and B., *Belemnites compressus*, Voltz., *Nautilus latidorsatus*, d'Orbig., *Turbo capitaneus*, Münst., *Pleurotomaria sub-decorata*, Münst., *Astarte lurida*, Sow., *Posidonomya Bronni*, Voltz., *Nucula Hausmanni*, Roem., *Gresslya gregaria*, Roem., *Lima bellula*, Mor. and Lyc., *Lima gigantea*, Sow., *Tancredia læviuscula*, Lyc., and several undescribed forms.

The Upper Lias at Illminster, Somerset, has become famous for the large number of species it has yielded to the long, patient, and careful investigations of Mr. Charles Moore, late resident in that locality. The beds here consist[1] of—

1. Rubbly beds 6—10 feet, containing *Ammonites communis*, Sow., *A. falcifer*, Sow., *A. Raquinianus*, d'Orbig., *A. bifrons*, Brug., *A. radians*, Rein., *A. insignis*, Schübl.
2. Clay, 8 inches.
3. Yellow limestone, 3—4 inches.
4. Layers of clay, 18 inches, *Leptæna Pearcei*, Dav.
5. Leptæna bed, 1 inch, *Leptæna Moorei*, Dav., *L. Bouchardii*, Dav., *L. liasina*, Bouchard.
6. Marlstone, 2½ inches, resting on greenish sand, containing Belemnites 4 inches and Marlstone.

Mr. Moore's museum contains *Teleosaurus*, *Ichthyosaurus*, and other reptiles, a magnificent collection of fishes, representing many new species of *Pachycormus*, *Lepidotus*, and other Upper Lias forms, in the finest possible preservation, the dark, enamelled scales of the fish contrasting finely with the pale yellow rock in which they are entombed. It was here likewise that Mr. Moore first discovered the Leptæna bed which contained

[1] Davidson's 'Monograph on British Oolitic and Liassic Brachiopoda,' Palæontographical Society, part 3, p. 17.

so many interesting forms of small Brachiopoda belonging to the genera *Leptæna*, *Spirifera*, *Thecidium*, *Rhynchonella*, and *Terebratula*, together with a number of *Gastropoda* and *Conchifera*, amounting to upwards of 150 species in all. The Leptæna bed forms a remarkable stratum, as it was found by M. E. Deslongchamps to occupy the same stratigraphical position at May, Calvados, as it does in Somersetshire and Gloucestershire. The following note, January, 1862, from my friend Thomas Davidson, Esq., explains the discovery of the Lias *Leptænas*: "When at Boulogne, in 1847, M. Bouchard received a parcel of fossils from the Lias of Pic de St. Loup, among which were several specimens of a small *Leptæna*, and about the same time I received a small parcel from Mr. C. Moore, among which I recognised two or three species of Lias *Leptæna*, which M. Bouchard and myself described, for the first time, in the 'Annals and Mag. of Natural History' for October, 1847. The discovery of the Lias *Leptænas* is therefore due to M. Bouchard and Mr. Moore. It was my description and publication of these species which first directed public attention to the subject."

List of Fossils from the Zone of Ammonites communis (*Gloucestershire*).

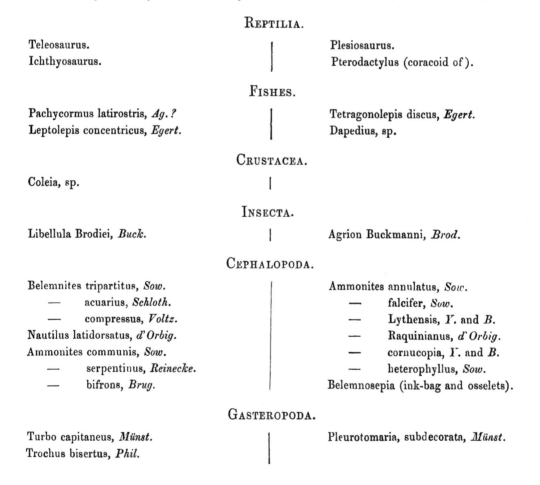

REPTILIA.

Teleosaurus.	Plesiosaurus.
Ichthyosaurus.	Pterodactylus (coracoid of).

FISHES.

Pachycormus latirostris, *Ag.?*	Tetragonolepis discus, *Egert.*
Leptolepis concentricus, *Egert.*	Dapedius, sp.

CRUSTACEA.

Coleia, sp.

INSECTA.

Libellula Brodiei, *Buck.*	Agrion Buckmanni, *Brod.*

CEPHALOPODA.

Belemnites tripartitus, *Sow.*	Ammonites annulatus, *Sow.*
—　　acuarius, *Schloth.*	—　　falcifer, *Sow.*
—　　compressus, *Voltz.*	—　　Lythensis, *Y.* and *B.*
Nautilus latidorsatus, *d'Orbig.*	—　　Raquinianus, *d'Orbig.*
Ammonites communis, *Sow.*	—　　cornucopia, *Y.* and *B.*
—　　serpentinus, *Reinecke.*	—　　heterophyllus, *Sow.*
—　　bifrons, *Brug.*	Belemnosepia (ink-bag and osselets).

GASTEROPODA.

Turbo capitaneus, *Münst.*	Pleurotomaria, subdecorata, *Münst.*
Trochus bisertus, *Phil.*	

Conchifera.

Astarte lurida, *Sow.*
Posidonomya Bronni, *Voltz.*
Nucula Hausmanni, *Roem.*
— ovum, *Sow.*
Gresslya gregaria, *Roem.*
Lima bellula, ? *Lyc.* and *Mor.*
— gigantea, ? *Sow.*

Tancredia læviuscula, *Lyc.*
Placunopsis sparsicostatus, *Lyc.*
Inoceramus dubius, *Sow.*
Monotis substriata, *Goldf.*
Arca inæquivalvis, *Goldf.*
Cucullæa Münster, *Ziet.*

Brachiopoda.

Leptæna Moorei, *Dav.*
— liasina, *Bouch.*
— granulosa, *Dav.*
Thecidium rusticum, *Moore.*
— Bouchardii, *Dav.*
Lingula Beanii, *Phil.*

Spirifer Ilminsterensis. *Dav.*
— Münsteri. *Dav.*
Rhynchonella pygmæa, *Mor.*
Terebratula globulina, *Dav.*
— Lycetti, *Dav.*

Echinodermata.

Acrosalenia crinifera, *Quenst.*
Pseudodiadema Moorei, *Wright.*

Pentacrinus.

The Zone of Ammonites Jurensis.

Synonyms.—" Lias Zeta," Quenstedt, ' Der Jura Uebersichtstafel,' p. 293. " Zone des *Amm. torulosus* und zone des *Amm. Jurensis*," Oppel, ' Juraformation,' p. 296. " Marnes d'Aresche, et Marnes de Pinperdu," Marcou, franc-comtois, les Roches, ' Du Jura,' p. 119, " 9ᵉ étage Toarcien (pars. sup.)," d'Orbigny, ' Cours. élément. de Paléontologie,' p. 469 (sect. at Thouars *h* to *l*). " Sands of the Inferior Oolite," Smith, De la Beche, Conybeare, and other English authors. " Cephalopoda bed and Upper Lias Sands," Wright, ' Quart. Jour. Geol. Soc.,' vol. xii, p. 292, 1856. " Ammonite Sands," Hull, ' Mem. of the Geol. Surv. Country around Cheltenham,' p. 25, 1857. " Cynocephala stage," Lycett, ' Cotteswold Hills,' p. 16, 1857. " Jurensis bed," Wright, ' Quart. Jour. Geol. Soc.,' vol. xiv, p. 25, 1857.

This youngest member of the great Lias formation has a limited geographical range; for like other unconsolidated arenaceous deposits, it has been extensively affected by denudation. It is well developed in the counties of Gloucester, Somerset and Dorset, and at Blue Wick, on the Yorkshire coast. It may be advantageously studied at the latter locality, and in the fine section at Frocester Hill, and in other smaller exposures in the Nailsworth and Brimscombe Valleys in Gloucestershire. The following section of

12

Frocester Hill, near Stonehouse, affords the best type of the zone of *Ammonites Jurensis.*

Section of Frocester Hill, near Stonehouse.

Fig. 30.

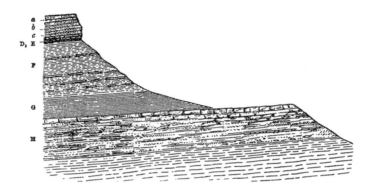

a, b, c. Inferior Oolite; 70 feet.

D, E. Calcareo-ferruginous sandstone (Cephalopoda bed); 6 feet.   ⎱ "Upper Lias Sands."

F. Yellow and brown sands, with inconstant and concretionary ⎰ Zone of *Ammonites Jurensis.*
    bands of calcareous sandstone; 150 feet?

G. Upper Lias shale; 80 feet = zone of *Ammonites communis.*

H. Marlstone; hard calcareous sandstone, resting on brown and gray sands, with bands and
    nodules of ferruginous sandstone; 150 feet = zone of *Ammonites margaritatus.*

I. Middle Lias shale = zone of *Ammonites capricornus.*

Inferior Oolite.

		Ft.	in.

a. A fine-grained oolitic limestone, similar to the freestones of Birdlip, Painswick, and
      Leckhampton Hills; the upper beds exhibit a most remarkable example of oblique
      bedding, the flaggy layers of which rest horizontally on inclined beds of freestone;
      thickness about 50   0

b. A coarse, light, cream-coloured, gritty, cystalline Oolite, traversed at intervals by ex-
      tremely crystalline shelly layers; a great part of the rock appears to be composed of
      fragments and plates of *Crinoideæ* plates and spines of *Echinidæ*, and comminuted
      fragments of the shells of *Mollusca*. This white rock has a most remarkable litholo-
      gical character, and glistens brilliantly when lit up by the sun's rays. The shelly
      and pisolitic seams which traverse this bed resemble those in the Pea-grit. The
      surface of weathered slabs exposes numerous microscopic objects; the rock, in fact,
      is almost entirely composed of organic debris, and measures about 10   0

c. A hard, fine-grained, oolitic, sandy limestone, of a light-brown colour, lithologically
      different from *b*. It contains many fossil shells, which are extracted with difficulty,
      and passes into a hard yellow Oolite with few fossils, attaining a thickness of
      from .... 8 to 10   0

[The lithological character of this rock is very different to that of *d*, on which it rests.]

The Cephalopoda Bed—Upper Lias.

ZONE OF AMMONITES JURENSIS.

Ft. in.

d. A coarse, dark-brown, calcareo-siliceous rock, full of small, dark, flattened grains of hydrate of iron. It contains an immense quantity of fossils, but *Ammonites* and *Belemnites* are the dominant forms; some of the bivalve shells are well preserved; the matrix adheres to the surface with such tenacity that they can seldom be cleaned without injury. The *Ammonites* and *Nautili*, for the most part, want the shell. *Rhynchonella cynocephala* lies in the upper part of the bed, and the *Ammonites*, *Belemnites, Nautili*, and other *Mollusca* in the middle part; the lower part is not so fossiliferous; this bed measures 4 6

e. A hard, coarse, brown mudstone, with hard irregular nodules of a calcareo-siliceous sandstone, highly micaceous and ferruginous, and passing downwards into the sands . 0 9

f. Fine, brown and yellowish, micaceous sands, passing into grayish coloured micaceous sands, with inconstant and concretionary bands of highly calcareous sandstone; nodules of various size occur in these bands, which are sometimes fossiliferous, containing chiefly *Ammonites* and *Belemnites*.. 150?

ZONE OF AMMONITES COMMUNIS.

g. Blue clay and shale, marked by the outburst of springs and by pools of water on the terrace formed by the Upper Lias Clay 80 0

ZONE OF AMMONITES MARGARITATUS.

h. Marlstone; a hard calcareous sandstone, resting on brown and gray sands, with bands and nodules of ferruginous sandstone 150 0

ZONE OF AMMONITES CAPRICORNUS.

i. The shales of the Middle and Lower Lias, sloping down into the valley,

Fossils of the Inferior Oolite.

A. Very few fossils in the Freestone; those observed were mostly fragmentary.
B. The fossils in this bed are so much broken that I have not been able to determine them. Stems and column-plates of *Extracrinus*, portions of the tests of *Pygaster semisulcatus*, Phil., and *Acrosalenia Lycetti*, Wr., plates of *Cidaris*, and quantities of spines in fragments, are seen on the slabs.
C. The following shells were observed, but could not be extracted from the upper part of the bed:

Pholadomya fidicula, *Sow.*	Trichites nodosus, *Lyc.*
Modiola plicata, *Sow.*	Serpula socialis, *Goldf.*

The frond of a Fern was found in this bed by the Rev. P. B. Brodie. The lower part of the rock resting on the Cephalopoda bed is sparingly fossiliferous.

In very few localities, where the sands are exposed along the escarpments of the Cotteswolds or in the beautiful valleys intersecting these hills, are they found to contain organic remains; fossiliferous veins have however been found at Frocester, Brimscombe, Nailsworth, Uley Bury, North Nibley and Ozleworth, and doubtless might be discovered in many other localities in this neighbourhood were the strata exposed.

The fossiliferous vein at Nailsworth lies near the base of the sands 4 or 5 feet above the Upper Lias clay. The bed consists of a fine soft ferruginous marly sandstone, of a deep brown colour, containing much peroxide of iron, and many shells, mostly of the same species as those found in the Cephalopoda bed at Frocester. The difference between these two beds is important, and deserves to be noticed, as the Cephalopoda bed at Frocester overlies the sands, whilst the fossiliferous vein at Nailsworth is found at their base, clearly proving that the sands and Cephalopoda bed form only one stage.

Fossils of the Zone of Ammonites Jurensis.

REPTILIA.

Vertebræ of *Ichthyosaurus*.

PISCES.

Teeth of *Hybodus*.

CEPHALOPODA.

Ammonites opalinus, *Reinecke*.
— Comensis, *von Buch.*
— insignis, *Schübler.*
— Aalensis, *Ziet.*
— hircinus, *Schloth.*
— Jurensis, *Zieten.*
— striatulus, *Sow.*
— complanatus, *Brug.*
— Thouarsensis, *d'Orb.*
— radians, *Reinecke.*
— striatulus, *Sow.*
— Moorei, *Lycett.*
— Boulbiensis, *Y.* and *B.*
— inornatus, *Williamson.*
— discoides, *Zieten.*

Ammonites Raquinianus, *d'Orb.*
— Levesquei, *d'Orb.*
— fimbriatus, *Sow.*
— Leckenbyi, *Lyc.*
— variabilis, *d'Orb.*, var. Beanii, *Simp.*
— variabilis, *d'Orb.*, var. dispansus, *Lyc.*
— obliquatus, *Y.* and *B.*, the aged form of variabilis.
Nautilus latidorsatus, *d'Orb.*
Belemnites compressus, *Voltz.*
— tripartitus, *Schloth.*
— irregularis, *Schloth.*
— Nodotianus, *d'Orb.*

GASTEROPODA.

Pleurotomaria subdecorata, *d'Orb.*
Chemnitzia lineata, *Sow.*
*Turbo capitaneus, *Münst.*

Trochus duplicatus, *Sow.*
*Natica adducta, *Phil.*
— Oppelensis, *Lyc.*

CONCHIFERA.

*Lima bellula, var., *Lyc.* and *Mor.*
*Modiola plicata, *Sow.*
*Perna rugosa, *Münst.*
*Hinnites abjectus, *Phil.*
*Pecten articulatus, *Goldf.*
*Gresslya abducta, *Phil.*
* — conformis, *Agass.*
  Myacites arenacea, *Lyc.*
*Homomya crassiuscula, *Lyc.*
*Goniomya angulifera, *Sow.*
*Myoconcha crassa, *Sow.*
*Cypricardia cordiformis, *Desh.*
*Pecten comatus, *Goldf.*
  Opis carinatus, *Wright.*
* — lunulatus, *Sow.*
  Cypricardia brevis, *Wright.*
  Cardium Hullii, *Wright.*
  — Oppelii, *Wright.*
  Cucullæa ferruginea, *Lyc.*

Cucullæa olivæformis, *Lyc.*
*Lima electra, *d'Orb.*
  Unicardium, nov. sp.
  Tancredia, nov. sp.
  Trigonia Ramsayii, *Wright.*
* — striata, *Sow.*
* — costata, *Sow.*
  Pecten textorius? *Goldf.*
*Pholadomya fidicula, *Sow.*
  — arenacea, *Lyc.*
  Lima ornata, *Lys. MS.*, nov. sp.
  Astarte lurida, *Sow.*
* — excavata, *Sow.*
  — detrita, *Goldf.*
  — complanata, *Roemer.*
  — rugulosa, *Lyc.*
  Gervillia fornicata, *Lyc. MS.*
* — Hartmanni, *Goldf.*
  Nucula Jurensis, *Quenst.*

BRACHIOPODA.

Terebratula subpunctata, *Dav.*

Rhynchonella cynocephala, *Rich.*
  — Jurensis, *Quenst.*

The species marked with an asterisk in the above list are found likewise in the Inferior Oolite; the specimens from the sands, however, are nearly all dwarfed forms, from which it is inferred that the physical conditions which then prevailed were unfavorable to their development. The stunted growth of the stationary *Conchifera* forms a striking contrast to the size, number and variety of the locomotive *Cephalopoda* interred with them in the same bed; in fact, the dawning existence of these *Conchifera* appears to have been a struggle for life, whilst the conditions under which the *Cephalopoda* existed were favorable to their continuance in time, as proved by the number of species and individuals of this class found in the Frocester beds; their life, notwithstanding, was abruptly brought to a termination by some great physical change which took place about the commencement of the deposition of the oolitic formations.

Leckhampton Hill, near Cheltenham, exhibits one of the most typical sections in Gloucestershire of the three sub-divisions of the Inferior Oolite, where the following beds are admirably exposed:—No. 1, 2 and 3 represent the zone of *Ammonites Parkinsoni*; No. 4 the zone of *Ammonites Humphriesianus*; No. 5, 6, and A, B, C, the zone of *Ammonites Murchisonæ*; these rest conformably on D, the Cephalopoda, or Jurensis bed, which is here very thin; E, F, G, is the Upper Lias resting on H, the Marlstone.

Section of Leckhampton Hill, near Cheltenham.

FIG. 31.

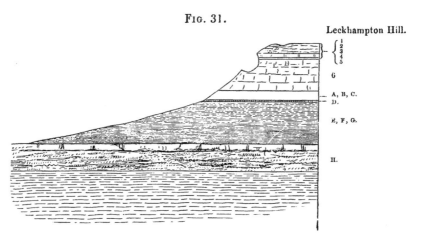

1. Trigonia bed.
2. Gryphæa bed.
3. Brown rubbly Oolite.
4. Flaggy freestone.
5. Fimbria bed or Oolite marl.
6. Freestone.

A, B, C. Pea-grit and ferruginous Oolite.
D. Cephalopoda or Jurensis zone.
E, F, G. Upper Lias sand and Upper Lias clay.
H. Marlstone.
I. Middle Lias clay-zone of *Ammonites capricornus*.

No. 1. The *Upper Trigonia bed* is a coarse brown ragstone, containing many fossils, chiefly as moulds and impressions of *Trigonia costata*, Sow., *T. decorata*, Lyc., *Lima cardiiformis*, Sow., *Rhynchonella concinna Terebratula spinosa*, Schl., Sow., *Ammonites Parkinsoni*, Sow., *Echinobrissus clunicularis*, Lhywdd, *Holectypus depressus*, Leske, and *Clypeus Plotii*, Klein; in thickness it is about seven feet.

No. 2. The *Gryphæa bed*, an ancient oyster bank, almost entirely composed of *Gryphæa sublobata*, Desh., with many other shells, as *Pholadomya Heraulti*, Ag., *Terebratula Meriani*, Opp., *Tancredia donaciformis*, Lyc., *Gervillia tortuosa*, Phil., and many other species; the dominant shell is the *Gryphæa*; this bed is about eight feet in thickness.

No. 3. The *Lower Trigonia bed*, a light-coloured, thin-bedded oolitic ragstone, containing a large assemblage of *Conchifera*, which in general have their shells preserved, with several species of *Echinodermata* and Corals.

No. 4. Upper flaggy bastard-freestone, well seen above the Oolite-marl: twenty-six feet thick. It represents the zone of *Ammonites Humphriesianus*; this rock is here almost non-fossiliferous, although the equivalent bed at Cleeve Hill contains a rich fauna.

No. 5. The *Fimbria bed* or Oolite marl, is a cream-coloured mud-stone, not unlike chalk-marl; the dominant shell is *Terebratula fimbria*, Sow.; it contains likewise *Lucina Wrighti*, Oppel., *Lima punctata*, Phil., *L. Pontonis*, Lyc., *Natica Leckhamptonensis*, Lyc., *Natica adducta*, Phil., *Mytilus pectinatus*, Sow., *Astarte elegans*, Sow., *Nerinæa*, sp., *Chemnitzia*, sp., and masses of Coral, chiefly *Thamnastræa Mettensis*, Edw. This bed was deposited under conditions very different to that of the freestone on which it rests; as its lower portion is slightly brecciated, and the surface of the freestone on which that breccia was deposited had been for some time exposed to aqueous action and made smooth thereby. The marl measures about seven feet in thickness, and passes upwards into a marly limestone, becoming oolitic in the uppermost layers. This division of the bed is about ten feet thick. The Fimbria bed is a constant feature in the Inferior Oolite of the Cheltenham district, and in the northern and middle Cotteswolds, but is absent in the southern parts of that range. It forms the upper part of the zone of *Ammonites Murchisonæ*.

No. 6. The *Freestone* is a compact light-coloured oolitic limestone; the uppermost beds are the best for building purposes; the middle beds are of an inferior quality, and are stained in part with the peroxide of iron; the lower beds contain large Oolite-grains, and are called "roestone;" the freestone in all is about 110 feet in thickness.

The Pea-grit (Zone of Ammonites Murchisonæ) Inferior Oolite.

		Ft.	in.
A.	A brown, coarse, rubbly Oolite, full of flattened concretions cemented together by a calcareous matrix. When the blocks weather, the concretions, which resemble flattened peas, form a very uneven surface. It contains many fossils in good preservation ...	12	0
B.	A hard, cream-coloured, pisolitic rock, made up of flattened concretions, with a thickness about similar to those in A.............	10	0
C.	A coarse, brown, ferruginous rock, composed of large oolitic grains; it is readily disintegrated by the frost, and is of little economical value. About.............	20	0

The Cephalopoda-bed (Zone of Ammonites Jurensis).

		Ft.	in.
D.	A brown marly rock, full of small dark oolitic grains of the hydrate of iron, which are strewed in profusion in a calcareous paste. About.............	2	0
D.	A thin seam of yellowish sand	0	1½
E.	A dark-gray crystalline limestone, extremely hard, and resembling some beds of the Carboniferous limestone; it is bored in different places by *Fistulana?*, the shells of which remain in the excavations	0	9
F.	A brown, argillaceous, sandy bed, full of micaceous particles; passing downwards into fine brown and yellow sands. Thickness unknown.		
G.	Upper Lias Clay, of a dark blue colour. Thickness probably	160	0

Fossils of the Pea-grit and Freestones.

CEPHALOPODA.

Ammonites Murchisonæ, *Sow.*
Nautilus truncatus, *Sow.*

Belemnites spinatus, *Quenst.*

GASTEROPODA.

Patella rugosa, *Sow.*
— inornata, *Lyc.*
Pileolus lævis, *Sow.*
Nerita costata, *Sow.*
— minuta, *Sow.*
Monodonta Lyelli, *d'Arch.*
— sulcosa, *d'Arch.*
Natica adducta, *Phil.*

Cirrus nodosus, *Sow.*
Trochotoma carinata, *Lyc.*
Turbo capitaneus, ? *Goldf.*
Trochus monilitectus, *Phil.*
Solarium Cotswoldiæ, *Lyc.*
Nerinæa cingenda, *Bronn.*
Actæonina Sedgvici, *Phil.*

CONCHIFERA.

Ostrea costata, *Sow.*
Placunopsis Jurensis, *Roem.*
Hinnites velatus, *Goldf.*
Limea duplicata, *Goldf.*
Lima sulcata, *Münst.*
— lyrata, *Münst.*
— Lycetti, Wright.
— bellula, *Mor. and Lyc.*
Pecten lens, ? *Sow.*
— Dewalquei, *Oppel.*
Mytilus furcatus, *Münst.*
— striatulus, *Goldf.*
Modiola Sowerbyana, *d'Orb.*
Avicula complicata, *Buck.*
Corbula involuta, *Goldf.*
Tancredia axiniformis, *Phil.*
Arca Prattii, *Mor. and Lyc.*

Arca pulchra, *Sow.*
— cancellata, *Phil.*
— lata, *Dunk.*
Trigonia costata (var. pulla), *Sow.*
— exigua, *Lyc.*
Astarte interlineata, *Lyc.*
— rhomboidalis, *Phil.*
Sphœra Madridi, *d'Arch.*
Cyprina trapeziformis, *Roem.*
Unicardium, nov. sp.
Myoconcha crassa, *Sow.*
Ceromya Bajociana, *d'Orb.*
Myopsis rotundata, *Buck.*
Cardium striatulum, *Phil.*
— lævigatum, *Lyc.*
Goniomya angulifera, *Sow.*
Pinna cuneata, *Bean.*

BRACHIOPODA.

Terebratula simplex, *Buck.*
— plicata, *Buck.*
— submaxillata, *Dav.*
Rhynchonella Wrightii, *Dav.*

Rhynchonella decorata, *Dav.*
— angulata, *Sow.*
— oolitica, *Dav.*
— nov. sp.

ANNELIDA.

Serpula grandis, *Goldf.*
— convoluta, *Goldf.*
— plicatilis, *Münst.*

Serpula quadrilatera, *Goldf.*
— flaccida, *Goldf.*

ECHINODERMATA.

Cidaris Fowleri, *Wright*.
— Bouchardii, *Wr*.
— Wrightii, *Desor*.
Rhabdocidaris Wrightii, *Desor*.
Acrosalenia Lycetti, *Wr*.
Pseudodiadema depressum, *Ag*.
Stomechinus germinans, *Phil*.
Polycyphus Deslongchampsii, *Wr*.
Pedina Bakeri, *Wr*.

Hemipedina tetragramma, *Wr*.
— perforata, *Wr*.
— Bonei, *Wr*.
Pygaster semisulcatus, *Phil*.
— conoideus, *Wr*.
Galeropygus agariciformis, *Forb*.
Goniaster obtusus, *Wr*.
Pentacrinus, nov. sp.

ANTHOZOA.

Montlivaltia Delabecheii, *Edw.* and *Haime*.
— Waterhousei, *E.* and *H*.
— cupuliformis, *E.* and *H*.
Axosmilia Wrightii, *E.* and *H*.
Latomeandra Flemingii, *E.* and *H*.

Isastræa tenuistriata, *E.* and *H*.
— limitata, E. and *H*.
Thamnastræa Mettensis, *E.* and *H*.
— Defranciana, *E.* and *H*.
— fungiformis, *E.* and *H*.

POLYZOA.

Stromatopora dichotomoides, *d'Orb*.
Diastopora Waltoni, *Haime*.
— Michelini, *Blainville*.
— Mettensis, *Haime*.
— Wrightii, *Haime*.
Spiropora straminea, *Phil*.

Lichenopora Phillipsii, *Haime*.
Neuropora damicornis, *Lamour*.
Heteropora conifera, *Lamour*.
— pustulosa, *Michel*.
Theonoa Bowerbankii, *Haime*.
Berenicea diluviana, *Lamour*.

13

FOSSIL ASTERIADÆ.

DESCRIPTION

OF THE

LIASSIC AND OOLITIC SPECIES.

Genus—URASTER, *Agassiz*, 1835.

STELLONIA, *Nardo*, 1834.
URASTER, *Agassiz*, pars, 1835.
ASTERIAS, *Grey*, 1841.
ASTERACANTHION, *Müller* and *Troschel*, 1840.
URASTER, *Forbes*, 1841.

Rays five, more or less cylindrical, and deeply cleft. The skeleton is composed of small, irregular-shaped ossicula, articulated together in a retiform manner, as seen in the subjoined section of a ray of *Uraster rubens*, Lin. (fig. 32) *a*, represents the long, femur-like, ambulacral ossicula; *b*, the small, short, inter-ambulacral ossicles; and *c* is the cavity in the ray produced by this arrangement.

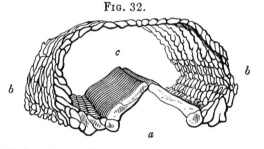

FIG. 32.

Section of a ray of *Uraster rubens* showing the arrangement of the calcareous ossicula.

The whole of the upper surface of the disc and rays is studded with blunt or pointed spines, either scattered singly or grouped together in tufts, and, whether single or fasciculate, arranged more or less regularly in longitudinal rows (Pl. I, fig. 2, *a*). The integument between the spines is naked, and shows the base of the spines; in the interspinous portions of the integument are many respiratory pores (Pl. I, fig. 2, *a*, *b*). Numerous pincers-like pedicellariæ, supported on soft stems are scattered among the spines or arranged in circles around their bases.

The wide ambulacral avenues are composed of two rows of long, compressed, femur-shaped bones, through which four series of tentaculæ or sucking-feet protrude (Pl. I, fig. 2, *b*). The vent is small and subcentral. The madreporiform body is single. The *Urasters* are found in all seas, but they prevail most in those of the Arctic and Atlantic regions; whilst in warm climates they are limited in numbers. Their presence, therefore, in any rock affords imperfect evidence of the climatal conditions under which it was deposited. All the oolitic species have hitherto been found in the Lias.

A.—*Species from the Lias.*

URASTER GAVEYI, *Forbes.* Pl. I, fig. 1, *a, b.*

> URASTER GAVEYI, *Forbes.* British Organic Remains, Memoirs of the Geological Survey, decade iii, plate ii, 1850.
> —   —   *Forbes*, in Morris's Catalogue of British Fossils, 2d ed., p. 90, 1854.
> —   —   *Wright.* Monograph of Brit. Oolitic Echinodermata, p. 428, 1855.

Rays five, moderately lanceolate; ambulacral areas wide and well exposed; ambulacral ossicles long, arcuate, and bi-carinate; sides and upper surface of the rays, closely covered, with short, tapering, thorn-like spines; the proportionate diameters of the disc to the rays is as one to six.

Dimensions.—Diameter of the disc, one inch and three twelfths; maximum breadth of a ray, eight twelfths of an inch; maximum breadth of an ambulacrum, five twelfths of an inch; length of a ray from the angle of junction with the disc, three inches and a half; length of an ambulacrum from its origin at the mouth, four inches.

Description.—The wonderful specimen figured in Plate I was obtained from the Middle Lias at Chipping Campden, in the Zone of *Ammonites capricornus;* the four rays which remain exhibit the anatomy of the skeleton in great perfection and disclose the close affinities it has with *Uraster rubens*, Lin., of our present seas, showing that this type of animal structure, at least, has undergone little modification during the inconceivable period of time which has elapsed since the Lias formation was deposited. This Star-fish lies on its upper surface, in a slab of Lias, among which are strewed in great abundance *Ammonites capricornus*, Schloth., *Unicardium cardiodes*, Phil., *Cardium truncatum*, Phil., *Cypricardia cucullata*, Goldf., and separate ossicles of *Pentacrinus robustus*, Wr. The under surface (fig. 1, *a*) is fully exposed, and small portions of the upper surface are likewise seen, which partly display the general character of the structure and clothing of the dorsal integument; the upper surface of the rays appears to have been covered with short, stout, tapering spines, set very closely together on that portion of the integument exposed on the slab; among these are scattered smaller and more slender spines with traces of *pedicellariæ ;* the wide ambulacral valleys are bordered by two marginal rows of thin plates, which lie obliquely on each other, with steep sides towards the valley, and having on their convex under surface four or five elevations with concave summits, to which the moveable marginal spines of the rays were articulated (fig. 1, *b*).

The wide ambulacral valleys are flattened, and of nearly the same breadth throughout, they taper a little towards the mouth and the end of the ray (fig. 1 *a*) ; the narrow depression down the centre indicates the suture by which the ambulacral ossicula were articulated along the mesial line of the ray; these bones (fig. 1,*b*) "are narrow and linear in shape, slightly bent, with the appearance of a very shallow sigmoid curve. This is caused by the curved keel which runs down each, grooved throughout two thirds of its length, but depressed and marked with two pit-like impressions in the neighbourhood of the ambulacral sulcus (fig. 1, *b*) ; the ends of the ossicula which go to form the sulcus are slightly denticulated. The curvation of the ossicula has reference to the disposition of the suckers, which in this genus are arranged in four series down each avenue. The perforations are slightly ovate in this species." Forbes. In figure 1, *b*, I have given an enlarged drawing of four of the ambulacral ossicula, and the corresponding bordering plate, with its mammillary articulating surfaces and spiny borders, for a comparison with the homologous parts of the ray in the living *Uraster rubens* (fig. 3).

Affinities and differences.—The structure of the ambulacral skeleton, which is so admirably preserved in this fossil, removes all doubt as to its true generic position and affinities. It approaches so much, in fact, the existing *Uraster rubens,* Lin., of our coasts, that it requires a careful comparison of its specific characters to determine the distinction which undoubtedly exists between this Star-fish of the Lias sea and that of our own time. It resembles in form *Uraster carinatus,* Wr., of the Marlstone of Yorkshire, but the prominent dorsal ridges in that species appear to be absent in *Uraster Gaveyi.*

Locality and Stratigraphical position.—This beautiful specimen was discovered by my friend, G. E. Gavey, Esq., F.G.S., in a slab of Middle Lias from Mickelton Tunnel, near Chipping Campden, Gloucestershire, on the Oxford, Worcester, and Wolverhampton Railway. The rock on which it lies belongs to the Zone of *Ammonites capricornus ;* with it are associated the following species of Echinoderms : *Cidaris Edwardsii,* Wr. ; *Hemipedina Bowerbankii,* Wr. ; *Tropidaster pectinatus,* Forb. ; *Palæocoma Gaveyi,* Wr. ; and *Pentacrinus robustus,* Wr.

URASTER CARINATUS, *Wright*, n. sp. Pl. II, fig. 1.

Rays five, long, and moderately lanceolate ; upper surface of the disc crowded with short, thick spines ; upper surface of the rays provided with three prominent carinæ, each formed of a series of stout conical spines, set closely together in rows, and inclined in an imbricated manner on each other.   The middle ridge is the broadest and most prominent.

Dimensions.—Diameter of the disc, two inches and a quarter; length of the only complete ray, from the intermediate angle of bifurcation, four inches and three quarters.

Affinities and differences.—In its general outline this Star-fish resembles *Uraster Gaveyi*, as we at present only know the under surface of that form and the upper surface of *Uraster carinatus* it is impossible to make a critical comparison of these Middle Lias Asteriadæ.

Locality and Stratigraphical Position.—This Star-fish was collected from the Marlstone at Bowlby, near Staithes, Yorkshire, and is the only example at present known.   It belongs to the rich collection of my friend, John Leckenby, Esq., F.G.S., of Scarborough.

Genus—TROPIDASTER,[1] *Forbes*, 1850.

Body stellate, five-rayed (a vent on the dorsal surface?); rays convex and carinated above, the carina composed of a double series of squamose plates; rest of the dorsal surface spinous; spines simple; ambulacra bordered by transverse plates, with spiniferous crests on their anterior margins; ambulacral ossicula rather broad, geniculated, pectinated at their inner extremities.   (Suckers biserial.)

A.—*Species from the Lias.*

TROPIDASTER PECTINATUS, *Forbes*.   Pl. III, figs. 1, 2, 3.

> TROPIDASTER PECTINATUS, *Forbes*.   Memoirs of the Geological Survey of Great Britain, Figures and descriptions of Organic Remains, Decade iii, pl. iii, 1850.
> —   —   *Forbes*, in Morris's Catalogue of British Fossils, 2d ed., p. 90.
> —   —   *Wright*. British Association Reports, vol. for 1856, p. 402.

Description.—My esteemed colleague, the late Professor Edward Forbes, gave so excellent an account of this Star-fish that I shall quote his description of the same entire. The general aspect of this species, when seen from above, is that of a *Uraster*, whilst viewed

[1] Τρωπις, *a keel*, and αστηρ, *a star*.

from below it resembles an *Astropecten*. The rays are rather short, about equal in length to the breadth of the disc, ovato-triangular. The upper surface of the rays and disc is covered with short, obtuse, simple spines, which, on the sides of the rays, are ranged in oblique rows of about five or six in each row. Placed rather laterally on the disc, is seen, though obscurely, a madreporiform plate, and I think I can perceive indications of an anal pore. Down the centre of each ray runs a keel composed of two rows of squamous plates, somewhat quadrate in form, but produced at their anterior and inner angles (fig. 2 *a*). This kind of keel, or mid-rib, is not present in any star-fish, recent or fossil, with which I am acquainted, and resembles in form and structure the tiling of the crest of the roof of a house. It must be regarded as a peculiarity of generic value. The rays themselves appear to have been very flexible, and not much liable to injury. Their extremities are rather pointed. On the under surface their centres are occupied by rather broad, lanceolate, ambulacral grooves running from the mouth (fig. 2 *a*) ; that orifice is somewhat contracted by the encroachment of the large, twin, triangular plates (fig. 3), with punctated surfaces, which occupy the angles formed by the junction of the bases of the rays. The ambulacral ossicula (fig. 2 *c*) are oblong, rather broad, strongly geniculated in the centre at their anterior margins, and denticulated by about five crenations at the edge which borders the ambulacral sulcus. The margins of the under surfaces of the rays are bordered by transverse, oblong, rather narrow plates (fig. 2 *c*), each bearing a crest at its anterior border, indented by the sockets of eight or more rather short cylindrical spines, which have rugose surfaces (fig. 2 *e*). It is these crested marginal plates, with their rows of spines, which give the under surface of this star-fish so much the aspect of an *Astropecten*.

Affinities and differences.— The affinities of *Tropidaster* have been so fully pointed out in the preceding description that it seems unnecessary to enter into more details on this branch of the subject. Since the specimen originally described by Professor Forbes was found, much larger and finer examples were discovered by my friend, Mr. Gavey, in the same locality; the best of these fossils I have figured in Pl. III, fig. 3, which shows very distinctly the large, prominent, twin, triangular plates around the mouth.

Locality and Stratigraphical Position.—This remarkable Star-fish was discovered by G. E. Gavey, Esq., F.G.S., in the Middle Lias at Mickelton Tunnel, near Chipping Campden, Gloucestershire, where it was associated with *Cidaris Edwardsii*, Wr., *Uraster Gaveyi*, Forb., *Palæocoma Gaveyi*, Wr., and *Pentacrinus robustus*, Wr., together with a series of Mollusca, characteristic of the zone of *Ammonites capricornus*, as *Ammonites capricornus*, Schloth., *Ammonites Henleyi*, Sow., and the usual species of Conchifera found interred with these Cephalopods.

Genus—SOLASTER, *Forbes.*

Crossaster, *Müller* and *Troschel.*

Body stellate; disc large, rays short and numerous; upper surface of the disc and rays covered with fasiculated spines; tegumentary membrane between the fasciculi naked; ambulacral furrows narrow, with two rows of pores for the tubular feet; no pedicellariæ; vent central.

Fig. 33.

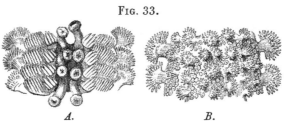

A.   *B.*

Portion of a ray of *Solaster papposa*, Linn.   *A*, the under; *B*, the upper surface.

The genus *Solaster* is represented in our epoch by only six species, two of which, *Solaster papposa*, Linn., and *Solaster endeca*, Linn., live in European seas. The only fossil which has hitherto been referred to this genus is the magnificent specimen figured in Plate IV, which was found at Windrush Quarry, Gloucestershire, in a block of light-coloured Oolitic freestone belonging to the Great Oolite; this unique fossil was obtained from the workman who discovered it by the Earl of Ducie, to whose collection it belongs.

A.—*Species from the Great Oolite.*

Solaster Moretonis, *Forbes.*   Pl. IV, fig. 1, *a, b, c, d, e.*

> Solaster Moretonis, *Forbes.*   Morris's Catalogue of British Fossils, p. 89, 2d ed., 1854.
> —   —   *Forbes.*   Memoirs of the Geological Survey, Organic Remains, Decade v, pl. i, 1856.
> —   —   *Wright.*   British Association Reports, vol. for 1856, p. 402.

Disc large; rays numerous, thirty-three in number, narrow, linear, equal lengthened, tapering to a fine point; ambulacral furrows wide and deep, the margins of the rays provided with several rows of fine, acicular, close-set spines.

Dimensions.—Diameter of the body, from ray point to ray point, five inches; diameter of the disc, one inch and four tenths of an inch; length of the rays, one inch and nine tenths of an inch; breadth of a ray at the widest part, four twelfths of an inch; average breadth of an ambulacrum, two twelfths of an inch.

Description.—This remarkable fossil Star-fish, as it lies on the slab, with its under side only exposed, has been likened to the head of a Crinoid, with outspread arms crushed flat, but the structure of the rays at a glance discloses its true characters. It was referred by Professor Forbes, who first described it, to the genus *Solaster*, as it has most affinities with that group ; the concealment of the dorsal tegumentary skeleton, however, prevents us from ascertaining with certainty whether it possessed paxillæ similar to those in existing species, and on which one of the main characters of the genus depends. In its general outline, and in the proportionate size of the disc, and the number and linear form of the arms, it resembles *Heliaster helianthus*, Linn., a many rayed species from the Pacific coasts of South America. A closer examination, however, of the structure of the rays shows that it belongs to a family in which the suckers are biserial, whereas all the *Urasteriadæ* have four rows of holes for the passage of tubular feet. It differs from all the living *Solasters* in having a proportionately smaller disc and a greater number of linear rays.

The skeleton of the disc is well preserved, it consists of a number of stout, oblong, rounded ossicles united together at their extremities, and forming a retiform structure, having a number of stellate centres at the junction of the ossicula, which are crowded together and overlap each other at these points ; (fig. 1 *e*) shows the reticular pattern of the ossicles of the disc magnified. The connecting ossicula only are seen, the spiniferous bones being concealed from view.

The skeleton of the rays has been so well described by my lamented colleague Professor Forbes, that I give the description in his own words. " The rays are very narrow and linear-shaped, their sides being parallel throughout the greater part of their length. At the point of junction of the base of each ray with that of the next, is a pair of erect, semi-circular, compressed, slightly sinuous, sharp-edged bones, the angle-ossicula (fig. 4 *d*), their inner edges, or those directed towards the mouth, approximate ; their outer edges are divergent. Their upper edges spread outwards, but much less so than in the corresponding bones in the recent *Solaster papposa*, Linn. ; and they are much more compressed and elevated. Along their outer margin are rows of slender spines which are admirably preserved in the specimen.

Each avenue is composed of two series of ambulacral ossicles, about sixty in a row, their inner edges being minutely crenulated and accurately meeting along the centro-sutural line (fig. 1 *b*). These ossicles are shaped something like a dice box, each divided into two more expanded portions and a central narrower part, (fig. 1 *b*). The inner portion is flattened or slightly excavated, and somewhat rhomboidal, the outer elevated into a ridge. The middle and more contracted portion is carinated obliquely, and on the inner (proximal) side has a triangular groove. A similar groove occurs on the outer (distal) side, placed nearer the middle than the former. The sides of the ossicles are widely excavated for the purpose of forming the ambulacral perforation through which the soft suckers or ambulacral feet passed. The inter-ambulacral ossicles are rather quadrate (fig. 1 *c*), and divided diagonally, though somewhat irregularly, and lobe-like, into two portions, of which the inner or inferior portion is elevated, and the outer depressed. These ossicles change

14

shape, and become narrower as they approach the buccal regions of the ventral disc. Their crests, or elevated portions, bear combs of long, slender, acicular spines, with bulbous bases; of these spines there are from four to six in each transverse row, (fig. 1 c) shows these quadrate spiniferous ossicula with their crests of comb-like spines magnified.

The arrangement of the dorsal surface of the rays is too obscure in the few portions of those organs that are reversed to enable one to make out their details with certainty; but I think I can perceive pretty clearly the paxillated character of the spines, and that these bodies, forming the radiated or brush-like crowns of the paxilla above described, are much shorter and stouter than the marginal spines.

Affinities and differences.—This fossil Starfish is quite unique; the organic characters of the skeleton so closely resemble those possessed by *Solaster papposa,* prepared expressly for the purpose of minute comparison, that I cannot doubt its being a true *Solaster,* the modifications in the form of the bones of the rays and in the number of these processes clearly prove, however, that it appertains to an extinct species, in which all the generic characters of the group are well preserved.

Locality and Stratigraphical position.—This Star-fish was discovered by the workmen at Windrush Quarry, Gloucestershire, in a block of oolitic freestone, belonging to the Great Oolite. It is now the property of the Earl of Ducie; the species was dedicated to his lordship by Professor Forbes, who first described it in the fifth decade of "British Organic Remains," published by the 'Geological Survey of Great Britain.'[1]

Genus—GONIASTER, *Agassiz.*

This genus, as established by Agassiz[2] in his Prodrome, includes Star-fishes with a pentagonal body, having the margin bordered with a pair of large plates, which sometimes carry spines; the upper surface of the body is covered with small tetragonal, or polygonal ossicles fitted within the marginal framework, the suckers are biserial, and the vent opens near the centre of the dorsal surface.

Müller and Troschel[3] suppressed the genus *Goniaster,* and formed, instead, three genera, *Astrogonium, Goniodiscus, Stellaster,* whose diagnostic characters were chiefly obtained from the structure of the marginal plates, as the following definitions indicate.

1. *Astrogonium.*—The large marginal plates are smooth towards the centre, and the border is surrounded by a circle of granules.

---

[1] 'Memoirs of the Geol. Survey, British Organic Remains,' pl. v, p. 3.

[2] "Prodrome d'une Monogr. des Radiaires Échinodermes," 'Mémoires de la Société des Sciences Naturelles de Neuchatel,' tome i, p. 191.

[3] 'System der Asteriden,' p. 52—62.

2. *Goniodiscus.*—The marginal plates have the whole of their upper surface granulated.

3. *Stellaster.*—The marginal plates are all granulated, and the ventral plates carry a suspended spine, as seen in the annexed figure, 34, of the under surface of a ray of *Stellaster Childreni.*[1]

FIG. 34.

Under surface of a ray of *Stellaster Childreni*, Grey.

For our present purpose, I retain the genus *Goniaster* as originally defined, inasmuch as the oolitic species hitherto discovered have not retained those delicate characters on which the sub-genera of the '*System der Asteriden*' were founded.

All the *Goniasters* have pentagonal bodies, with five angles, indicating the extremities of the rays, which in some species project more or less. The disc is always flat in dried specimens, or when removed from the water, but is capable of considerable elevation in their native element. They however always want the con-vexity of *Asteropsis* and the elevation of *Oreaster*. All the species have their margins bounded by two rows of large marginal plates (fig. 35 *c*), which enter into the formation of the sides of the disc and arms, and are always larger than the discal plates which occupy the upper and under surfaces of the body (fig. 35 *b*), and (fig. 36 *A* and *B*).

FIG. 35.

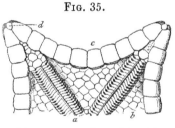

Astrogonium cuspidatum, M. and T.

The size and characters of the marginal plates render them, in a palæontological point of view, the most important parts of the skeleton, as they are almost always well preserved, and afford, at the same time, characters which appear to be very constant in the different species. Their surface is either smooth or granulated, surrounded by granules or without decoration, some having spines or pedicellariæ, others being without such appendages. The marginal plates enter into the formation of the border and form a firm frame-work, into which all the other parts appear to be fitted (fig. 36). Among the living species, the character of the marginal plates and their appendages is apparently much more constant than among the fossil forms, and for this reason they have been used by zoologists for classificatory purposes.

[1] See p. 47, for further details of these genera.

The upper and under surfaces of the disc are covered with a kind of Mosaic composed

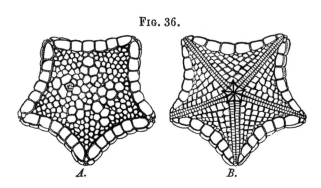

FIG. 36.

A. *B.*

of small tetragonal or polygonal ossicles united by a delicate membrane; these plates are not firmly articulated together, and they are consequently easily displaced, so that they are rarely preserved in a fossil state. The ossicula of the disc and margin usually support small granular spines, and sometimes sessile pedicellariæ. The ambulacral furrows are bordered by square ossicula, and the surfaces are marked by deep parallel grooves which serve for lodging the ambulacral spines (fig. 35 *a*). The marginal plates towards the termination of the rays are modified and enlarged for supporting and protecting the eyes (fig. 35 *d*).

A.—*Species from the Inferior Oolite.*

GONIASTER OBTUSUS, *Wright.* Pl. II, fig. 3 *a*, *b*, 3 *c*, 3 *d*.

GONIASTER OBTUSUS, *Wright.* British Association Reports, vol. for 1856, p. 402.

Rays elongated, rounded at the extremity; marginal plates convex and prominent, the upper row larger than the under, the entire surface of both plates covered with fine granulations; a single row of small polygonal plates between the upper marginals, the ocular plate at the end of the ray prominent.

The fragment above described is the largest portion of a *Goniaster* I have collected from the Pea Grit, Inferior Oolite. I have frequently obtained individual bones belonging to the marginal skeleton of this genus, but the conditions under which this deposit was accumulated appear to have been unfavorable for the conservation of the entire body. This fragment proves, however, that the GONIASTERIADÆ appeared with the dawn of the Jurassic epoch along with many new forms of Echinodermata in the seas that deposited the first oolitic sediments.

I found associated with this Star-fish *Cidaris Fowleri*, Wr., *C. Bouchardii*, Wr., *C. Wrightii*, Des., *Stomechinus germinans*, Phil., *Pseudodiadema depressum*, Ag., *Polycyphus Deslongchampsii*, Wr., *Pedina Bakeri*, Wr., *Hemipedina perforata*, Wr., together with *Galeropygus agariciformis*, Forb., and *Pygaster semisulcatus*, Phil., in considerable numbers.

B.—*Species from the Great Oolite.*

GONIASTER HAMPTONENSIS, *Wright.* Pl. II, fig. 2.

GONIASTER HAMPTONENSIS, *Wright.* British Association Reports, vol. for 1856, p. 402.

Body pentagonal, sides arched, rays projecting in the form of cones, and tapering to a point; marginal plates thick, sides elevated, and inclined inwards; upper surface of the disc covered with small polygonal ossicles.

Dimensions.—Diameter of the body from ray point to ray point, three inches; diameter of the disc, from the inner side of one superior marginal plate to the same point of the opposite margin, one inch and a quarter; depth of the border at the centre of the arched side, three-tenths of an inch.

Description.—The specimen figured in Pl. II, fig. 2, is the only *Goniaster* which has been found at Minchinhampton. It is unfortunately broken, and the portion preserved is so much incorporated with the matrix, that the sculpture on nearly all the marginal plates is destroyed. In the most perfect side, there are twenty-five upper marginal plates, but the lower series cannot be counted. The margin is high, and inclined inwards (fig. 2 *b*). The five rays project like narrow cones from the sides of the disc, thereby producing the arching of the margin so characteristic of this species (fig. 2 *a*).

The upper surface of the disc was covered with small tetragonal or polygonal plates, which have been so much effaced in carving out the fossil from the oolitic matrix, that sections only of a few of them remain. On one or two upper marginal plates, I have seen a finely granulated surface, whilst all the others are pitted by the oolitic grains during the process of cystallization in the replacement of the test.

Affinities and differences.—This oolitic species very much resembles some cretaceous forms, as *Goniaster Smithii* and *Goniaster Coombii;* the form and structure of the marginal plates, and the clothing of the upper disc appear very similar in both, there is no other oolitic species at present sufficiently known with which it can be compared; the mere fragment of *Goniaster obtusus,* Wr., from the Inferior Oolite, does not afford materials for comparison. The fine-pointed termination of the rays in *Goniaster Hamptonensis,* Wr., however, is very different from the blunted termination of the ray in *Goniaster obtusus,* Wr.

Locality and Stratigraphical position.—This specimen was discovered by Mr. Edward Day, many years ago, in the planking beds of the Great Oolite of Minchinhampton Common, by whom it was cleared or rather carved out of the soft freestone in which it was imbedded;

and sold by him to Professor Buckman, to whose collection it now belongs. In fig. 2 *a*, the fossil is represented lying on the under side, having the upper surface fully exposed; in fig. 2 *b*, a lateral view of the same is given, for the purpose of showing the height and inclination of the marginal plates; in neither figure has Mr. Bone ventured to delineate the ossicles of the disc.

FAMILY—ASTROPECTENIDÆ, *Müller* and *Troschel*.

The species of this family have a stellate body flattened on both sides. The rays are narrow, elongated, and bordered by one or two rows of marginal ossicula. The ventral plates are always spiniferous, and the dorsal, when present, are covered with granules which are more or less so likewise. The narrow ambulacral valleys, have two rows of holes for the passage of tubular feet. The upper surface of the body, between the marginal plates, is covered with paxillæ closely set together, and the under surface is crowded with short spines arranged in regular rows.

In the classification of Müller and Troschel this family comprises the genera *Astropecten* and *Ctenodiscus*, which possess a double series of marginal plates, and *Luidia*, with only a single ventral row of marginal spiniferous ossicula. At (page 48) the reader will find a diagnosis of the genera of this family; to these I have added the genus *Plumaster*, an extinct form from the Lias which has many affinities with *Luidia*.

Genus—LUIDIA, *Forbes*.

Rays elongated, and numerous. Margin provided with a single row of plates, instead of a double row as in *Astropecten*. These ventral marginal ossicula carry spines, as seen in the subjoined figure of the section of a ray of *Luidia Senegalensis*, where A shows the upper, and B the lower surface. The upper surface of the body is covered with small close-set paxillæ (A). The ambulacral valleys are narrow, and the suckers biserial (B).

FIG. 37.

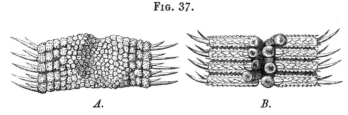

A.   *B.*

Portion of a ray of *Luidia Senegalensis*, M. and T. *A*, the upper; *B*, the under surface.

Two sets of spines are found on the underside of the ray, the central portion has rows of short, stout spines, and the marginal plates are armed with long recurved spines.

A.—*Species from the Lias.*

LUIDIA MURCHISONI, *Williamson.* Pl. V, fig. 2.

LUIDIA MURCHISONI, *Williamson.* Magazine of Nat. History, vol. ix, p. 425, 1836.
— MURCHISONI, *Forbes.* Mem. of the Geol. Surv., vol. ii, part 2, p. 480.
LUIDIA MURCHISONI, *Forbes*, in Morris's Catalogue of British Fossils, 2d ed., p. 83.
SOLASTER POLYNEMIA, *Simpson.* Fossils of the Yorkshire Lias, p. 135, 1855.
LUIDIA MURCHISONI, *Wright.* British Association Report, vol. for 1856, p. 402.

Rays twenty, moderate in length, obtuse at the apex, and having their margins fringed with numerous short, hair-like spines, mouth opening large, with impressions of radial processes.

Description.—This unique fossil Star-fish is so imperfectly preserved that only a few of its characters can be ascertained. It was first figured and described by Professor Williamson, in the ninth volume of Loudon's 'Magazine of Natural History' for 1836, and was thus described:

"This fossil was found in the marlstone at the point where it is carried up into the cliff to the north of the great fault, at the Peak Hill near Robin Hood's Bay, near the lower part of the stratum, where it blends with the lower lias. The slab on which the fossil is preserved is of a rather micaceous nature, a matrix, generally unfavorable for preserving minute characters; and a portion of the fossil having adhered to the upper part of the rock which fell in pieces, the view presented is rather that of the internal than the external structure of the animal. The central circle, the situation of the mouth, is preserved very distinctly, and proceeding with considerable regularity from this, is a series of rays, twenty in number. Those rays near their base bear the sulcus (furrow) which runs under those of recent Asteriæ; but towards their apex they become more worn and thin, showing in several places a small wiry line, with short ribs branching off at right angles, apparently a species of appendage, resembling what represents the vertebral column and ribs in the turtle, and which is observable in recent Asteriæ. There are also slight traces of transverse grooves on the whole surface of each ray; but these are generally almost obliterated. Along the margins are extremely regular rows of small rhomboidal perforations, or cells, from which proceed a series of lateral filaments, or delicate lengthened papillæ; but on the surface of the fossil, it merely presenting to us the interior, no papillæ are preserved. The apex of such rays as have not been broken off prior to the animal being entombed, are obtusely pointed."

Locality and Stratigraphical position.—From the appearance of the shale in which this specimen is embedded, it appears to come from the zone of *Ammonites capricornus*, it therefore belongs to the Middle Lias, and occupies about the same horizon as *Uraster Gaveyi*, Wr., from the Middle Lias of Gloucestershire.

Genus—PLUMASTER, *Wright.*   1861.

Rays numerous, long, and plume-like ; narrow at the base, expanded in the middle, and tapering gently towards an obtuse apex. The inter-ambulacral ossicula are much elongated transversely, they have a row of spiniferous tubercles on the middle of their under side, and their outer distal margin is pectinated fig. 1 *b.* ; the tubercles carry long hair-like spines, the ambulacral ossicula are thick and prominent, like the vertebral bones of *Ophiuriadæ ;* the avenues are narrow and depressed. The radial bones, at the base of the rays, form a conspicuous prominent ring around the mouth opening (fig. 1 *c*).

A.—*Species from the Lias.*

Plumaster ophiuroides, *Wright.*   Pl. V, figs. 1 *a*, 1 *b*.

Rays twelve, three times as long as the diameter of the disc, bent, and plume-like ; narrow at the base, expanded in the middle, with obtuse terminations ; ambulacra narrow and depressed, forming a furrow in the centre of the ray ; lateral ossicula long, bent, and slightly arched, with a row of tubercles on the centre of each bone. The proximal side of the ossicula, in relation to the disc, is slightly convex, and the distal side pectinated on its outer half, (fig. 1 *b*). The rows of tubercles support long spines which lie *in situ* in the specimen. The radial bones around the disc circle are very prominent, and resemble ancient trusses with a sculptured surface (fig. 1 *c*).

Affinities and differences.—This Star-fish resembles *Luidia* in the general form of the rays, whilst it differs from that genus in the structure of the lateral ossicles, and the possession of a conspicuous row of tubercles along the centre of each bone. It has some resemblance to *Pteraster,* but the fringe of marginal spines which forms so remarkable a character in *Pteraster* is wanting in *Plumaster.* These characters may be said to be generic, rather than specific, and as the specimen under consideration is a unicum, little more can be positively stated on the subject.

Locality and Stratigraphical position.—This beautiful Star-fish was found by Mr. Peter Cullen in the shales of the Middle Lias near Skinningrave Bay, on the Yorkshire coast; many of the rays are well preserved, the anatomical details however are partly concealed by an irremoveable pyritic film. This unique specimen is the property of my friend, John Leckenby, Esq., F.G.S., and forms one of the many rarities contained in his rich cabinet of Yorkshire fossils.

Genus—ASTROPECTEN, *Linck.*, 1733.

STELLARIA, *Nardo*, 1834.
ASTERIAS, *Agassiz*, 1835.
—   *Forbes*, 1841.
ASTROPECTEN, *Müller* and *Troschel*, 1842.

Body stellate, flat on both sides, rays elongated. Two rows of large marginal plates at the border. The lower series provided with spine-like scales, which increase in size from

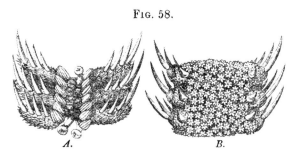

FIG. 58.

within outwards, and terminate in long, moveable spines. The dorsal marginal plates are covered with granules, which often become spinous, and sometimes carry spines. The flat upper surface of the body and rays is thickly covered with appendages, the summits of which are crowned with groups of minute spines. This genus is the most abundant in the oolitic Rocks, the Lias, Inferior Oolite, Great Oolite, Kelloway Rock, Coral rag, Kimmeridge

Portion of a ray of *Astropecten polyacanthus*, M. and T. *A*, under surface; *B*, the upper surface of the ray.

clay, and Portland beds. All contain species characteristic of each of those divisions of the Jurassic series; the structural characters of these fossils are so admirably preserved in all our examples that we have no difficulty in referring them to the existing genus *Astropecten.*

A.—*Species from the Lias.*

ASTROPECTEN HASTINGIÆ, *Forbes.* Pl. VI, fig. 3 *a, b,* fig. 4 *a, b.*

ASTROPECTEN HASTINGIÆ, *Forbes.* Memoirs of the Geological Survey of Great Britain, vol. ii, part 2, p. 478, 1848.
—   —   *Forbes.* Ibid., Brit. Organic Remains, decade 1st, pl. ii, fig. 1.
—   —   *Wright.* British Association Reports, vol. for 1856, p. 402.

Rays five, short, acute, lanceolate, sides straight, intermediate angles obtuse; marginal plates quadrate, subequal; surface of the disc, on the upper and under sides, covered with small, tetragonal ossicles, arranged in a tesselated order.

Dimensions.—Diameter of the body from ray point to ray point, nearly two inches; diameter of the disc, one third of the whole.

Description.—Rays short in proportion to the rather broad, flat body, triangularly lanceolate, with very straight sides and pointed extremities. The angles formed by their junction with each other and the body are obtuse. Their margins are bordered by regular series of nearly equal square plates, decreasing but slightly as they approach the apex. The length of each ray is about equal to the diameter of the disc. There are about eighteen marginal plates in each row. The surface is covered by quadrate tesselations, indicating the arrangement of the plates, which probably, when the animal was alive, bore tufts of paxillæ or coronated spines. The specimen measures two inches in diameter.

Affinities and differences.—*Astropecten Phillipsii*, Forb., is probably its nearest fossil ally, but the form and characters of its surface distinguish it conspicuously from any other British member of its genus.

Locality and Stratigraphical position.—This species was discovered in the marlstone of Yorkshire; the original specimen, which I have not seen, nor can I discover in whose possession it now is, was figured and described by Professor Forbes from an example in the late Marchioness of Hastings' collection; I have therefore adopted entire my lamented colleague's description of this specimen. The other example (Pl. VI, fig. 4 *a, b*) I found on a slab of marlstone, from Boulby, near Staithes, associated with *Uraster carinatus,* Wr.; this specimen belongs to my friend, John Leckenby, Esq., F.G.S., and is now in his cabinet. Fig. 4 *a* shows the under surface of the Star-fish, and fig. 4 *b* the angular ossicles near the base of the rays; fig. 4 *b*, the ambulacral valleys, with the biserial tentacule pores and numerous quadrate plates, arranged in a tesselated manner.

B.—*Species from the Inferior Oolite.*

ASTROPECTEN LECKENBYI, *Wright.* Pl. VII, fig. 1 *a, b, c.*

Rays five, elongated, borders straight, intermediate angles very obtuse, border thick; marginal plates quadrate, elongated transversely, surface closely covered with large granules; madreporiform tubercle large, button-shaped, opposite one of the intermediate angles.

Description.—The border of this *Astropecten* is formed of thick marginal plates, of a quadrate form, the transverse diameter of which is twice as much as the length; their surface is covered with large granules, arranged in a quincuncial order, as shown in fig. 1 *c.* The rays are all more or less broken, consequently the entire number of marginal plates cannot be ascertained; in the most perfect ray remaining there are thirty-four plates. As the under surface of this Star-fish is immoveably fixed to the matrix, its upper surface alone is exposed (fig. 1 *b*); the thickness of the marginal plates is well shown in section here. The remains of the large madreporiform body occupies an excentral position, on the upper surface, opposite one of the intermediate angles (fig. 1 *a*); as none of the structure of the intermarginal surface of the body is preserved, the character of the paxillæ is unknown.

Locality and Stratigraphical position.—This Star-fish was discovered in a boulder of Gray Limestone on the shore, by the White Nab, near Scarborough. The rock was described by Prof. Phillips as Great Oolite, but the zoological character of its fossils proves that the species are identical with the fauna from the zone of *Ammonites Humphriesianus* of the Inferior Oolite, to which I have referred the Gray Limestone in my memoir.[1]

Astropecten Leckenbyi was here associated with *Ammonites Humphriesianus*, Sow., *A. Blagdeni*, Sow., *A. Brackenbridgii*, Sow., and *A. Parkinsoni*, Sow., together with twenty species of Gasteropoda and forty species of Conchifera, all of which are, for the most part, characteristic of the Inferior Oolite. *Rhabdocidaris maxima*, *Pseudodiadema depressum*, Ag., *P. vagans*, Phil., *Astropecten Scarburgensis*, and *Ophiura Murravii*, Forb., are likewise found with it in the same bed of Gray Limestone.

This beautiful specimen is now preserved in the Scarborough Museum, and is the only example of the species that has yet been found.

ASTROPECTEN SCARBURGENSIS, *Wright.*   Pl. VII, fig. 2 *a, b, c.*

Rays five, elongated, tapering to an acute point, intermediate angles acute; border of the rays convex; marginal plates thick, quadrate; surface covered with small, close-set irregular granules.

Dimensions.—Transverse diameter from ray point to ray point, three and a half inches; diameter of the disc across the marginal plates, one inch.

Description.—In many respects this species resembles *A. Leckenbyi*, but it differs from that form in having the intermediate angles more acute, the borders of the rays convex instead of straight, and the surface of the marginal plates covered with small irregular granules. As both these Star-fishes are unicums, I write with much reserve regarding their differences, knowing how often external characters are found to blend into each other when they are

---

[1] 'On the Subdivisions of the Inferior Oolite in the South of England,' and 'Quart. Jour. Geol. Soc.,' vol. xvi, p. 29.

examined and compared with the various modifications of form and structure which a number of specimens of the same species often exhibit after an attentive study thereof.

The rays in *Astropecten Scarburgensis* taper to an acute point (fig. 2 *a*), the borders are slightly convex, and there are about fifty ossicles around the margin of each ray; the ossicula are quadrate, much rounded, and have their surface covered with numerous small, close-set, granules (fig. 2 *c*); the ambulacral valleys are wide, and the ambulacral bones are elongated and quadrate (fig. 2 *b*); the upper and under marginal plates form well-marked prominences on the border, when examined in profile, as shown in this figure; a few small discal ossicles occupy the intermarginal spaces of the rays.

Locality and Stratigraphical position.—This specimen was discovered by Mr. Peter Cullen in the Gray Limestone near Scarborough. It belongs to the cabinet of my friend John Leckenby, Esq., F.G.S.

C.—*Species from the Stonesfield Slate.*

ASTROPECTEN COTTESWOLDIÆ, *Buck.* Pl. IX, fig. 3 *a, b, c,* fig. 4 ; Pl. X, fig. 1 *a, b, c, d,* and fig. 3 *a, b, c, d.*

ASTERIAS COTTESWOLDIÆ, *Buckman.* Murchison, Geol. of Cheltenham, 2nd ed., tab. iii, fig. 5, p. 94, 1845.
ASTROPECTEN COTTESWOLDIÆ, *Forbes.* Memoirs of the Geol. Survey, p. 479, 1848.
— — *Forbes.* Morris's Catalogue of British Fossils, 2nd ed., p. 72, 1854.
— — *Wright.* British Association Reports, vol. for 1856, p. 402.

Body flat, rays five, elongated, tapering to an acute point, border of the rays straight, intermediate angles obtuse ; dorsal marginal plates quadrate, upper surface convex, and covered with small granules ; ventral marginal plates spiniferous on their posterior borders ; upper surface of the disc provided with five oblong, bilobed eminences, having two internal rows of serrated processes ; a series of similar bilobed bones, diminishing in size towards the apex, occupy the middle of the rays, with which the discal eminences appear to be continuous; madreporiform tubercle large, excentral, with fine radiating lamellæ.

Dimensions.—Diameter of the body, from three to four inches ; the proportionate diameter of the disc to that of the body is as one to three and a half.

Description.—This beautiful Star-fish is sometimes found in fine preservation on slabs of Stonesfield slate. The flat body is provided with five elongated rays, which gradually taper to an acute termination ; the border of the ray is quite straight, and contains fifty-five to sixty dorsal marginal plates, which have a quadrate form, being rather broader than long; they are well separated from each other, by reason of the convexity of their upper surface; the inner side of each ossicle is flat, the sides straight, and the outer margin and upper surface rounded, which imparts a moniliform character to the border of the ray; the convex surface of the ossicles is crowded with minute granules, most conspicuous near the vertex and disappearing on the sides.

The ventral border plates (Pl. X, fig. 1 a, b, c, d) have a rectilineal arrangement, and gradually diminish in size from the angle to the apex; on their posterior border is a row of three or four little tubercles, which support short, stout spines (Pl. X, fig. 1 c and d), and many of these are preserved *in situ;* in the only specimen in which I have seen the ventral surface exposed, these spines form a well-defined row, projecting outwards and backwards from the distal border of the ventral plates (Pl. X, fig. 1 b, c, d.)

The upper surface of the disc in the specimen figured in Pl. IX, fig. 1 a, exhibits five prominent, oblong bodies, bilobed in structure, and having on their inner surface a number of tooth-like processes, which mutually interlock. At fig. 1 b I have given an enlarged view of these bodies; they appear to be the upper portion of the large ambu-lacral bones, for we observe a series of similar bodies occupying the middle of the rays in other specimens where the matrix has been cleared away from these ossicles, as seen in each of the five rays of the specimens figured in Pl. IX, fig. 1 a, fig. 1 c; and in another specimen from the same locality (Pl. X, fig. 3 b, fig. 3 d) the ambulacral bones are well seen; the serrated line between the two halves of the oblong ossicles is the suture at the middle of the arch formed by the ambulacral bones; the position and relation of this median suture may be understood by referring to the section of a ray of *Uraster rubens* which I have given in fig. 32 a, page 99.

The madreporiform tubercle is moderately large, and situated near the border opposite one of the intermediate angles; it consists of numerous fine, vertical laminæ, radiating from the centre, and which so closely resemble the septa of a *Montlivaltia* that the tubercle might readily be mistaken for a small fossil coral attached to the upper surface of the Star-fish.

The ventral surface (Pl. X, fig. 1 a) is shown in the only specimen I have seen with the under side exposed; the mouth-opening (fig. 1 a), is surrounded by five petals, each formed of two halves, and perforated in the middle for the passage of a tube, as shown in the enlarged drawing (fig. 1 b); at each intermediate angle there is an arch of small ossicles, two and three rows deep, arranged in an imbricated manner, and over the summit of each arch the point of one of the five petals rests; this arrangement produces a curious complicated structure, which is very well delineated in fig. 2 b; the surface of all the little ossicles entering into the formation of these arches is covered with small, spiniferous tubercles.

Within the ventral border plates are two rows of small, quadrate, spiniferous ossicles, forming the outer walls of the narrow ambulacral furrows, the inner border of these ossicles supports numerous rows of small spines, as shown in fig. 1 *b*, and in the enlarged fig. 1 *c* the margin of the furrow is seen to have a comb-like structure where it closes over the ambulacral groove; the same figure shows likewise rows of stout, thorn-like spines projecting outwards from the ventral border plates.

Affinities and differences.—This Star-fish resembles very much *Astropecten Wittsii*, Wr., from the same beds; it has, however, a proportionately larger disc, more obtuse intermediate angles, and the rays more attenuated at their termination, as shown in the figures given of the two species in Pl. IX, figs. 1, 2, 3.

It has many points of affinity with *Astropecten Phillipsii*, Forb., but as I have not seen the original specimen of Prof. Phillips's drawing, I can offer no remarks on that form. Unless the actual specimens can be compared, figures scarcely afford data sufficiently accurate on minute points of structure, to justify critical observations on the affinities and differences existing between species nearly allied to each other.

Locality and Stratigraphical position.—This beautiful Star-fish was first discovered in Gloucestershire, by the Rev. E. F. Witts, in slabs of Stonesfield slate at Eyeford, near Naunton, and was figured by Prof. Buckman in the second edition of Sir R. Murchison's 'Geology of Cheltenham.' Good specimens, showing the marginal plates in relief, are rare, whilst sections of the skeleton are not uncommon in this fissile limestone, associated with the teeth of Fishes, the elytra of Insects, and the shells of Mollusca.

ASTROPECTEN COTTESWOLDIÆ, *var.* STAMFORDENSIS, *Wright.* Pl. VI, fig. 1 *a*, *b*.

Samuel Sharp, Esq., discovered a beautiful Star-fish in a slate bed near Stamford, which appears to be the equivalent of the Stonesfield slate. I have figured this specimen in Pl. VI, fig. 1, under the name *Astropecten Cotteswoldiæ*, var. *Stamfordensis*, Wr., as it presents some few traits of structure which distinguish it from the Eyeford specimens of the Cotteswold Hills. The upper marginal bones of the Stamford specimen are more angular, the intermediate angles more acute, and the borders of the rays straighter. As the specimen is a unicum, a critical comparison between forms so nearly allied is exceedingly difficult unless more of the anatomy of the Star-fish was exposed than happens to be in Mr. Sharp's specimen; I therefore prefer considering it provisionally as a variety of *Astropecten Cotteswoldiæ* rather than risk its separation on imperfect and it may be transient characters from that well-marked form already described in detail. I am indebted to Mr. Sharp for a description of the rock and locality where he found the specimen, and have much pleasure in adding his letter, of date 27th April, 1859, which he sent me in reply to inquiries made relative to the exact stratigraphical position of the bed from whence he obtained this Star-fish, and which leaves no doubt of its being true Stonesfield slate, even if the lithological character of the matrix in which the fossil lies had not of itself been sufficiently characteristic of that formation.

" The Star-fish forwarded to Mr. Bone was found by me on the 13th of April, 1853, in the parish of St. Martin, in Stamford, which parish stands in Northamptonshire, separated from the other parts of the town by the river Welland, which at this point divides the counties of Northampton and Lincoln. There is a very curious fact in connection with the beds upon which Stamford and St. Martin's stand. The valley of the Welland, which separates the old town from its outlying parish, narrows at this point almost to a gorge, and the ground rises steeply on either side. On the northern or Lincolnshire side the surface rises through and at the back of the town to an elevation of from 150 to 200 feet above the level of the river, and the strata of which this hill is composed consist of the Cornbrash; capping the higher elevation of the district is a thick bed of variegated and stratified clays equivalent to the Bradford clay. A series of beds of the Great Oolite, attaining a great thickness, and having as their lowest member a slaty bed occasionally very fossiliferous, and proved, I believe, to be equivalent to the Stonefield slate of Oxford-shire. Beneath these are the white, siliceous sand of the Inferior Oolite, and the Oolitic rock of the same formation, strongly impregnated with iron. Under all, and forming the basement bed of the district, lies the Lias clay, of unascertained thickness. The beds I have enumerated preserve, as nearly as may be, their horizontal position; they are intersected more or less by fissures, but otherwise exhibit little evidence of disturbance. Upon the Northamptonshire side, however, an upheaval has taken place, throwing up the lower beds to the top of the hill, and disturbing their horizontal position. The upper beds of the series are wanting on this side, but a huge fragment seems to have separated from the upheaved mass, and to have subsided into the chasm formed by the convulsion.

" Upon this fragment St. Martin stands; it is about three quarters of a mile in length, east and west, by about half a mile in width, north and south; its beds preserve nearly their original horizontal position, but they are divided near their western extremity by a fault running north and south.

" Thus, in proceeding from the river southward (up the hill), we have the Lias clay at the bottom, then the ironstone of the Inferior Oolite and its overlying sands, then the *slate bed*—Stonesfield slate—and over this some of the very various beds of the Great Oolite. Still rising the hill, we come again to the Lias clay, the ironstone, the sands, slates, and Great Oolite, in a reiterated succession; but in this second series we have what we had not before observed, a bed of clay lying between the beds of the Great Oolite and the beds of the Inferior Oolite, and which clay, from its position, I suppose to be equivalent to the Fuller's earth.

" It was in the *slate bed* of the *lower* series above described that I found the *Astropecten* in what is locally called a " pot-lid." The lower beds of slates consist of masses of a flattened, semi-spherical form, lying with their convex surfaces downwards on the underlying bed of sand, in the hollows of which, it would seem as if they had been moulded.

" The exact geological position in which the specimen was found, is thus precisely fixed as at the base of the Great Oolite beds and the summit of the Inferior Oolite beds, in a stratum equivalent to and, in its general characteristics, I believe, identical with the *Stonesfield slate* of Oxfordshire."

ASTROPECTEN WITTSII, *Wright*.   Pl. IX, fig. 2 *a, b.*

> ASTROPECTEN WITTSII, *Wrght.*   British Association Reports, vol. for 1856, p. 402.
> —   —   *Wright.*   Monograph of Oolitic Echinodermata, p. 428, Pal.
> Soc., 1858.

Body flat; rays five, elongated, tapering to a blunt termination, border bulging at the middle of the ray, intermediate angles acute, disc small in proportion to the body; dorsal marginal plates quadrate, elongated transversely; upper surface of the disc and arms between the border plates, convex and prominent.

Dimensions.—Diameter of the body from ray point to ray point, two and a half inches; diameter of the disc from the outer side of one angle to the opposite, eleven twentieths of an inch, extreme breadth of a ray, three tenths of an inch.

Description.—This Star-fish resembles *Astropecten Cotteswoldiæ* in its general structure, but differs from it so much in the proportionate smallness of the disc to the diameter of the body and in the bulging of the rays that I have separated it from that more common form. It may probably be found to be a variety of *A. Cotteswoldiæ* should connecting links between the two forms hereafter be discovered; in the mean time, however, I have separated it from that species and dedicated it to my friend, the Rev. E. F. Witts, who first discovered Star-fishes in the Stonesfield slate of the Cotteswold Hills.   The disc is small in proportion to the diameter of the body, in the ratio of one to five; the rays, five in number, are nearly of the same width throughout; the borders bulge slightly in the middle, and they terminate in blunted extremities; this is very apparent in the upper ray of fig. 2 *a*, but still more so in the specimen now before me.   The marginal plates are quadrate, little elongated transversely; the surface is covered with fine granules, and there are about fifty plates around the border of the most perfect ray; the intermarginal space is convex and prominent, and the dorsal integument, with its numerous paxillæ, appears to be well preserved; in this specimen (fig. 2 *b*), the disc exhibits five elevations, corresponding to the inner ambulacral bones, within which it is slightly depressed.   The remains of the madreporiform body, represented in the enlarged drawing (fig. 2 *b*), is seen close to the angle of the base of the longest ray; the intermediate angles are very acute, and the border plates compactly arranged, without the moniliform appearance seen in some allied species.

Affinities and differences.—This species very much resembles *Astropecten Cotteswoldiæ,* it differs from that form, however, in having the disc smaller in proportion to the diameter of the body, the rays a little swollen out in the middle, with their terminal extremities blunt and not attenuated, as in *Astropecten Cotteswoldiæ ;* these characters may, perhaps, belong to varieties of that species, for the remark already made in reference to *A. Cotteswoldiæ* var. *Stamfordensis,* is applicable to this and other unicums, that were a number of different forms of the same species before us it is possible we might be able to link together differences by a series of gradations, without which the extreme varieties of typical forms might be considered characteristic of different species, for it is doubtless true that each species has its own limits of variation ; in some these are circumscribed, in others they are enlarged, and it is by observation alone that the boundary between varieties and specific forms can be ascertained ; hence the difficulty which surrounds the investigations of the palæontologist, for his materials are in general fragmentary, in many cases unique, and always requiring the most careful study, critical comparison, and accurate analysis ; so that, without inclining to Darwinian notions on the one side or to modern species-making on the other, we feel the necessity of the greatest caution in pronouncing on specific differences between forms of which we only know solitary examples, and of these but partial details.

Locality and Stratigraphical Position.—This Starfish was collected from the Stonesfield slate of Eyeford, near Naunton, Gloucestershire, by the Rev. E. F. Witts, of Upper Slaughter, to whom I have dedicated the species, as an acknowledgment of his original discovery of *Astropecten Cotteswoldiæ* in the oolitic rocks of Gloucestershire.

ASTROPECTEN COTTESWOLDIÆ, *var.* STONESFIELDENSES. Pl. VIII, fig. 2.

Marginal plates thick and prominent, fifty around each ray, border straight, intermediate angles obtuse, inter-marginal spaces of the disc and rays covered with small ossicles; the plates have become so crystalline and weathered that all· their delicate sculpture and other characters are effaced.

Dimensions.—Diameter of the body from ray point to ray point, four inches ; diameter of the disc, one inch and one eighth.

This specimen belongs to the British Museum, it was bought at the sale of the late Mr. Johnson, of Bristol, and is supposed to be from the Stonesfield slate of Oxfordshire. It appears to be a large example of *Astropecten Cotteswoldiæ,* but the condition of the skeleton renders any minute examination thereof impossible.

D.—*Species from the Forest Marble.*

ASTROPECTEN PHILLIPSII, *Forbes.*   Pl. X, fig. 2, *a, b, c, d, e.*

> ASTROPECTEN PHILLIPSII, *Forbes.*   Memoirs of the Geological Survey of Great Britain, vol. vi, part 2, p. 478, 1848.
> —   —   *Forbes.*   Ibid., British Organic Remains, decade 1st, pl. ii, fig. 2, 1849.
> --   —   *Wright.*   British Association Reports, vol. for 1856, p. 402.

Rays five, elongate, lanceolate; margins straight; intermediate angles obtuse; marginal plates quadrate, and transversely elongated, surface tuberculated, the tubercles supporting short, stout spines.

Dimensions.—Diameter of the body from ray point to ray point, about five inches; diameter of the disc, one inch and one sixth; breadth of a ray at the base six tenths of an inch. These measurements are only approximate, as the specimen is slightly distorted.

Description.—Disc moderately developed, the arms being in length, compared with its diameter, as one and three quarters to one.

Rays slender, lanceolate, forming very obtuse angles at the junction with each other and the body. Margins of the rays bordered with oblong quadrate plates, which are studded with small tubercles, probably marking the points of attachment of paxillæ; on their edges also are a few scattered linear-lanceolate spines, which are not equal to the breadth of the plate. The ambulacra are bordered with semicircular combs of short spines.

The plates composing the skeleton of the body appear to have been oblong. The marginal plates at the angles are narrow, as compared with those of the ray borders. The diameter of the body is about one inch and one sixth. The length of the rays appear to have been about two and one sixth inches; their breadth, near the junction of the rays with the body, is about seven twelfths.

Affinities and differences.—This beautiful species bears a striking resemblance to the recent *Astropecten arantiacus*, Lin., and its allies. No fossil species of this genus, as yet figured, so clearly proves the true generic position of the extinct forms as this.

Locality and Stratigraphical Position.—This lithograph was executed from a drawing by Prof. Phillips of a specimen obtained from the upper sandy beds of Forest Marble at Hinton-lane-end, Yorkshire; fig. 2 *a* shows the under surface; fig. 2 *b* the structure and arrangement of the ambulacral bones; fig. 2 *c*, the surface of one of the marginal plates, magnified, with the small tubercles on its surface; fig. 2 *d*, the mode of articulation of the marginal plates with the row of stout spines arming their posterior border; fig. 2 *e*, one of these spines magnified, and showing its articulation to the tubercle.

ASTROPECTEN PHILLIPSII? Pl. X, *a*, fig. 2.

Rays five, short, tapering to an acute point; border thick and quite straight; intermediate angles obtuse; marginal plates moderately large, nearly quadrate, about fifty around the border of each ray; ventral plates only exposed, the outer border of each armed with short, stout, thorn-like spines; ventral portion of the disc wide; ambulacral furrows broad.

Dimensions.—Breadth of the disc, one inch and a quarter; diameter of the body, from ray point to ray point, three and a half inches; proportionate diameter of the disc to the length of a ray, one and a quarter to one and a half inches.

This Star-fish was figured in 'Loudon's Magazine of Natural History' for 1829, vol. ii, p. 73, and was thus noticed by a Yeovil correspondent, August 21st, 1828 :— " I send you a drawing of the *Fossil Asteria* found at Horsington, by the Rev. James Hooper, Rector of Stawell. It was taken from a stratum of Cornbrash, and is a very perfect specimen. The sketch and the figure is of the exact size of the original." Being most anxious to obtain the original specimen, in order to give a better drawing of this beautiful Star-fish, with details of its structure, I commissioned a friend to make inquiries in the neighbourhood about the fossil, for since the discovery of the specimen Mr. Hooper had died, and his family had left. The collection, I learned, had been taken to Ireland, and I regret to add that I have been unable to trace the specimen. I am, therefore, under the necessity of reproducing the original figure from Loudon, more with the view of inducing geologists living near Yeovil to search the Cornbrash of that locality for other specimens of this Star-fish than for any scientific value in the figure itself. I have provisionally referred this species to *Astropecten Phillipsii*, as it resembles that form more than any other yet discovered.

ASTROPECTEN HUXLEYI, *Wright*, nov. sp. Pl. VIII, fig. 1 *a*, *b*, *c*, *d*.

Rays five, broad, gently bent, with sloping borders; intermediate angles obtuse; marginal plates quadrate, elongated transversely; surface covered with small, flattened tubercles, those on the ventral series supporting short, spatulate spines; several, longer, thorn-like spines project from their posterior border; ambulacral furrows wide; the small inter-ambulacral bones at their border carry tufts of small, spatulate, equal-sized spines, which form a boundary for the ambulacral furrows; mouth surrounded by five ridge-like terminations; ventral integument covered with numerous rows of elongated, spatulate spines.

Dimensions.—Diameter of the ventral disc, one inch and eight tenths; length of a ray, about two inches and three tenths.

Description.—This Star-fish has all the five rays gently bent, indicating a flexible condition of the articulation of the marginal bones; this bending of the rays is well delineated in the drawing; the everted ray exhibiting the same curvature, which proves that the bending of the rays was dependent on the form of the articulating surfaces of the marginal bones, and not on muscular contraction, seeing that it was persistent after death.

The surface of the marginal plates, forty in number in the portion of the longest ray remaining (and probably the part which is absent had twenty more), is closely covered with small, flattened tubercles; those on the ventral plates support short, spatulate spines, which clothe the entire surface of the plates in an imbricated manner (fig. 1 *c*); the dorsal marginal plates are likewise covered with similar flattened tubercles, and many of the small spines which they supported are seen *in situ;* at the distal border of the ventral series several of the longer, thorn-like spines are seen which armed the border of the rays (fig. 1 *d*).

The everted ray exhibits in a very satisfactory manner the structure of its under surface; the ambulacral furrow is wide, and bounded by a series of small inter-ambulacral bones (fig. 1 *c*), which carry combs of small, flat spines, with flattened terminations on their lower edges. Fig. 1 *c* shows these tufts of spines hanging like festoons of fringes along the border of the ambulacral furrows (fig. 1 *b* and fig. 1 *c*). The inter-marginal tegumentary membrane of the disc and dorsal surface of the rays was strengthened with small ossicles, and the remains of the paxillæ are strewed abundantly amongst these calcareous pieces.

The tegumentary membrane on the ventral surface of the disc was covered with long, spatulate spines, which were disposed in rows in an imbricated manner; the under surface of the rays had likewise ranges of short, flat spines, set in rows on the marginal plates, and combs of spines on the inter-ambulacral bones, so that the under surface of this Star-fish was everywhere clothed with small, spatulate spines; the distal border of the ventral marginal plates had a row of larger, thorn-like spines projecting outwards.

The ventral surface of the disc is wide, and around its centre are five ridge-like bodies, with small tubercles on their surface (fig. 1 *a*); these bony processes might have been employed as jaws.

Affinities and differences.—This species is nearly allied to *Astrospecten Phillipsii,* Forb., but the greater size of the disc, with the breadth and curved shape of the rays, distinguish these forms from each other. The structure of the marginal plates and the inter-ambulacral bones, with their combs of spatulate spines, afford additional evidence that these two Star-fishes appertain to distinct species.

Locality and Stratigraphical Position.—This Star-fish which belongs to the British Museum, was collected from the Forest Marble near Malmesbury, by Mr. William Buy, where it was associated with numerous Mollusca, Echinidæ, and Crinoidæ, appertaining to that formation.

I dedicate this species to my friend Professor Huxley, F.R.S., whose numerous contributions to palæontology and zoology have advanced the progress of these branchés of natural science.

E.—*Species from the Kelloway Rock.*

ASTROPECTEN CLAVÆFORMIS, *Wright.* Pl. XI.

> ASTERIAS ARENICOLA, *Charlesworth.* London Geol. Journ., tab. 17, 1847.
> ASTROPECTEN ARENICOLUS, *Forbes.* Mem. of the Geol. Surv., vol. ii, part 2, p. 477, 1848.
> —   —   *Forbes.* In Morris's Catalogue of British Fossils, 2nd ed., p. 72, 1854.
> —   CLAVÆFORMIS, *Wright.* Mongr. of Oolitic Echinoderms, Pal. Soc., p. 428, 1860.

Rays five, convex, contracted at the base, enlarged at the inner third by the breadth of the marginal ossicula, and tapering gently throughout the two outer thirds of their length; disc small, its proportionate diameter to the body as 2 to 9; intermediate angles acute, and much contracted by the bulging of the rays; marginal plates variable in width, transversely elongated, and in general exceeding one third of the ray. Ambulacral furrows straight and linear, not participating in the enlargement of the margin, which is entirely owing to the form and development of the border plates.

Dimensions.—Diameter of the disc from one angle to another, one inch and eight tenths; diameter of the body, nine inches; width of a ray near the angle, one inch; width of a ray at the widest part of its enlargement, one inch and one fifth; from its maximum width to its worm-like point it gradually tapers.

Description.—This Star-fish was first figured in the 3rd part, pl. 17, of the 'London Geological Journal,' by Mr. Charlesworth, under the name of *Asterias arenicola*, Goldf., from the belief that it was identical with that species; in this opinion Professor Forbes[1] concurred, for we find in his memoir on ' British Fossil Asteriadæ' the following diagnosis of this form under the name *Astropecten arenicolus*, Goldfuss:

[1] Memoirs of the 'Geological Survey,' vol. ii, part 2, p. 477.

"*A. radiis* lanceolatis, longis, acuminatis, ad origines contractis ; angulis intermediis acutis ; *ossiculis marginalibus* angulorum brevibus, in parte latiori radiorum maximis, angusté oblongis, in apicibus radiorum quadratis.

"This species," he adds, " measures nearly a foot in diameter. The peculiar form of the rays, which, united by their bases at an acute angle (where the marginal plates are the narrowest), then swell out into a petaloid shape, and again contract into long, linear-lanceolate extremities, distinguishes it from all congeners. Each ray is to the diameter of the disc as three to one. There are about seventy plates on each side of each ray.

"Marlstone of Yorkshire. It was first described and figured from the Oolites of Germany."

A comparison of our figure with that of Goldfuss, shows that the German Star-fish although allied to, is specifically distinct from, our fossil ; the disc is larger in proportion to the body, the area of the rays is likewise greater, and the marginal bones are wider, with a different order of increase ; the intermediate angles are likewise more obtuse ; there is no contraction at the base ; the general form of the ray is entirely different, tapering to an obtuse termination, and having the marginal bones widest at the outer third of the ray ; whereas in *Astropecten clavæformis*, Wr., the intermediate angles are very acute where the marginal plates are narrowest ; they then suddenly swell out to their maximum breadth, and having attained that width, they gradually and regularly diminish, terminating in fine, worm-like extremities. These comparative differences clearly prove that *Astropecten clavæformis* constitutes a well-marked species, entirely distinct from *Astropecten arenicolus*, Goldf., with which it has hitherto been identified.

Astropecten clavæformis, Wr., is always found in the form of moulds, having the external figure of the body well preserved ; many of these retain tolerably sharp impressions of the different external characters of the species ; from one of these moulds, my friend, Mr. C. R. Bone, obtained a beautiful cast in gutta percha, showing the general contour of the body ; aided by this and the impressions on the moulds, he has been enabled to produce the very truthful figure of this ancient Star-fish which he has given in Plate XI.

By the same process we have been enabled to figure, in Plate X A, fig. 3, a remarkable four-rayed variety of this species which I found in the Museum of the Yorkshire Philosophical Society. This specimen shows that the *Asteriadæ* of the oolitic fauna were liable to deformities of the same character as are found so frequently to prevail among their congeners of the present day.

Affinities and differences.—The nearest affinity of *Astropecten clavæformis* is with *Astropecten arenicolus*, Gold. ; the most obvious points of difference I have pointed out in the preceding paragraph. It is readily distinguished from *Astropecten rectus*, McCoy, in which the border of the rays is perfectly straight, without any enlargement near the base. Prof. Forbes described *Astropecten Orion* as a very regularly stellate species, having gradually tapering arms, bordered by square plates, which decrease regularly and gradually towards the apices. This species was likewise extremely spiniferous, and possessed numerous rows of spines on the margin of the ambulacral avenues, which are absent in *Astropecten clavæformis*.

Locality and Stratigraphical Position.—Some strange blunders have been made about the rock in which this species is found. Prof. Forbes, in the memoir already quoted, calls it the Marlstone of Yorkshire, and Prof. Morris, in his 'Catalogue of British Fossils,' states it to come from the Lias of Yorkshire; local collectors nearly all refer it to the Calcareous Grit, and it is so catalogued in the York and other Museums. These, however, are mistakes, as the sandstone from which this Star-fish is obtained belongs to the Kelloway rock, which is well seen in position resting on the Cornbrash in a quarry near the Leavisham station on the Whitby branch of the North-Eastern Railway. The specimens are always in the form of moulds of the exterior, and in no instance has a fragment of any of the ossicula been discovered.

ASTROSPECTEN ORION, *Forbes.* Pl. X, fig. 1 *a.*

ASTROPECTEN ORION, *Forbes.* Memoirs of the Geol. Survey, vol. ii, part 2, p. 478, 1848.
— — *Forbes.* In Morris's Catalogue of British Fossils, 2nd ed., p. 72, 1854.
— — *Wright.* British Association Reports, vol. for 1856, p. 402, 1856.
— — *Wright.* British Oolitic Echinodermata, Palæont. Society, p. 428, 1860.

Rays five, linear, lanceolate, tapering to a blunt point; border straight, intermediate angles obtuse marginal plates small, quadrate, numerous, 55—60 around the border of a ray, each plate carries numerous small spines; disc moderately large; ambulacral valleys wide, bounded by one or more rows of short, stout, thickly set spines.

Dimensions.—Diameter of the disc, one inch and three quarters; length of a ray, three inches; length across the fossil, from ray point to ray point, six inches. This specimen is small and immature, and one ray is broken; a specimen, in the British Museum, measures eight inches across.

Description.—This Star-fish is found only as a mould in a sandstone bed of the Kelloway rock; most of the specimens, beyond the mere outline of the fossil, have few details preserved. In the museum of the Yorkshire Philosophical Society, however, I found a specimen in which the characters of the marginal plates and spines were well preserved; this the council of the society kindly communicated for this work. From this mould Mr. Bone obtained a very good cast in gutta percha, and from the relief and the specimen he has been able to produce, the excellent figure given in Pl. X A, fig. 1.

The marginal plates are small, square ossicles, which decrease in size gradually and regularly towards the apex of the rays; in the specimen figured, which is small, there are from 55—60 marginal plates around the border of the most perfect ray, and in larger specimens, where the number of the border plates increase with age, they may amount in a full-grown individual of this species to 80. These marginal plates are very spiniferous; in addition to the row of thorn-like spines at their distal border, the dorsal plates appear to have been clothed with small spines, from the numerous impressions they have left in the mould, and which are reproduced in the cast. The border is quite straight, and the rays taper gradually and regularly from the base to the apex.

The ambulacral valleys are wide and well defined; the small bones bounding these avenues supported one or more rows of short, stout spines, the numerous impressions of which are well preserved in the specimen and have been admirably reproduced in the cast.

As Prof. Forbes gave no figure of *Astropecten Orion*, nor marked by that name any specimen in the museum in Jermyn Street (then under his care), considerable doubt exists as to the Star-fish he had described as *A. Orion*. After the best consideration I have been able to give to the subject, I have come to the conclusion that the fossil I have figured must be the species intended. In the 'Memoirs of the Geological Survey' Prof. Forbes states that all the specimens of *A. Orion* are from the Oolites of Yorkshire, but in the list of Echinodermata published in Prof. Morris's 'Catalogue of British Fossils,' and supplied by Prof. Forbes, this species is said to have been obtained from the Lias of Yorkshire, and the same error is committed regarding *A. clavæformis*, which is associated with *A. Orion* in the same formation; in fact, at the time my esteemed colleague published his notes on British Fossil Asteriadæ, the true position of the sandstone containing these moulds of Star-fishes was not known, for by some the rock was called Calcareous Grit and by others Marlstone, whereas, in fact, it is a bed of the Kelloway rock. *Astropecten Orion* is thus described by Prof. Forbes:

"*A. radiis* lineari-lanceolatis, longis, lateribus rectis, angulis intermediis obtusis; *ossiculis marginalibus* omnibus (ossiculis angulorum exceptis) plus-minus ve quadratis spiniferis.

"Measures eight or more inches in diameter. A very regularly stellate species, having gradually tapering arms, bordered by square plates, which decrease regularly and gradually towards the apices. Each ray is to the diameter of the disc as three and a half to one. There are about forty ossicula on each side of each ray.

"In the collections of the Marchioness of Hastings, the Marquis of Northampton, and Dr. Bowerbank. All the specimens are from the Oolites of Yorkshire."

Affinities and differences.—This species resembles *Astropecten Phillipsii*, Forb., in its general outline and leading characters, but differs from that form in having the marginal ossicles smaller, more numerous, and more spiniferous; the crowded rows of spines which bound the ambulacral valleys form a remarkable character in this Star-fish.

It differs from *Astropecten clavæformis*, Wr., associated with it in the same formation, in having the border straight, the marginal ossicles small, square, and spiniferous, and in having the rays lanceolate and tapering to a blunt apex. The swelling out of the ray towards the base, and the form and size of the marginal plates in that region, constitute diagnostic characters by which the two forms are readily distinguished from each other.

Locality and Stratigraphical Position.—All the specimens at present known have been obtained in the state of moulds from a bed of light-coloured sandstone appertaining to the Kelloway rock, near Leavisham station, on the Whitby branch of the North-Eastern Railway. In Newton Dale there is a fine development of the Middle Oolites, and most instructive sections of Kelloway rock, Oxford Clay, and Calcareous Grit are well exposed on each side of this railway which takes the line of the highly picturesque valley of the Esk.

The specimen I have figured belongs to the museum of the Yorkshire Philosophical Society. The British Museum, the Museum of Practical Geology, Jermyn Street, London, the Scarborough Museum, and the cabinets of Dr. Murray and John Leckenby, Esq., F.G.S., Scarborough, all contain fine specimens of this Star-fish.

F.—*Species from the Calcareous Grit.*

ASTROPECTEN RECTUS, *McCoy.* Pl. XII.

ASTROPECTEN RECTA, *M'Coy.* Ann. and Mag. of Nat. Hist., 2nd series, vol. ii, p 408, 1848.
— RECTUS, *Forbes.* Morris's Catalogue of British Fossils, 2nd ed., p. 73, 1854.
— — *Wright.* British Association Reports for 1856, p. 402, 1856.
— — *Wright.* Monograph of Oolitic Echinodermata, p. 428, 1858.

Rays five, narrow, elongated, tapering gradually from the angles to the apex, sides straight, bordered by two rows of quadrate marginal plates, averaging in width one third of the ray; surface covered with small tubercles and a row of larger, spiniferous tubercles on the distal border of each ossicle; intermediate angles acute; disc small in proportion to the body, proportionate diameter as two to nine.

Dimensions.—Diameter of the disc, one inch and nine tenths; breadth of the body from ray point to ray point, nine inches; length of a ray from the intermediate angle to the point, four inches and one quarter; breadth of a ray at the widest part, nine tenths of an inch.

17

Description.—This species is characterised by its long, linear, narrow, straight-sided rays, bordered by two rows of large marginal, quadrate ossicula, much lengthened transversely, each bone being about one third the length of the diameter of a ray; the whole of their convex upper surface is covered with small tubercles, as shown in our figure; at the posterior side of each bone is a row of three or four larger tubercles, which supported large spines; there are about ninety marginal bones around the border of a ray; along the inner side of the marginal plates a series of smaller quadrate ossicles is placed, three of these equalling the length of two of the marginal bones; the ambulacral ossicula are likewise covered with minute tubercles, and on the centre of each bone is one large tubercle for supporting a gigantic spine.

This Star-fish is for the most part known to us by horizontal sections of the skeleton as figured in Pl. XII. In one instance only have I seen the marginal bones free from the matrix, and these are figured in the same plate; this specimen which formerly belonged to the collection of Mr. Bean, of Scarborough, is now contained in the York Museum.

The diameter of the disc to that of the body is as two to nine, and many specimens attain a foot in diameter. The hard crystalline character of the Calcareous Grit in which this fossil is found conceals all details of the external structure of the ossicula beyond those I have figured.

Affinities and differences.—This species is well characterised by the narrowness and straightness of the rays and the large size of the marginal plates. It differs, therefore, from *Astropecten clavæformis* in the absence of the enlargement at the inner third of the rays which constitutes so remarkable a character of that form. Count Münster, in his 'Beiträge zur Petrefacten-Kunde' figured in pl. xi, fig. 1, of that work, a Star-fish under the name *Asterias Mandelslohi* obtained from the sandstone of the Inferior Oolite near Aalen, which has some resemblance to *Astropecten rectus*, but the marginal plates are proportionately narrower, and the ambulacral furrows considerably wider in the Aalen fossil than in our species.

Locality and Stratigraphical Position.—This species is obtained by splitting open the large nodules which fall out of the Calcareous Grit under Filey cliffs, near Filey Brig, on the Yorkshire coast. Mr. Peter Cullen informed me that all his specimens were obtained from this locality; they are, however, merely horizontal sections of the body and arms, for in only one specimen have I seen the form and surface of the marginal plates, I obtained a very large specimen of this species from the Calcareous Grit near Calne, in Wiltshire, which measures from the centre of the disc to the end of the ray six inches, giving the dimension of one foot to the body.

A MONOGRAPH

ON THE

FOSSIL ECHINODERMATA

OF THE

OOLITIC FORMATIONS.

THE OPHIUROIDEA.

The Ophiuroidea were long united with the true Star-fishes, Asteroidea, from the circumstance that in both groups the arms proceed from the circumference of the disk; they are now, however, separated into a distinct order, as they possess certain permanent organic characters by which they are distinguished from them.

The Ophiuroidea have a central discoidal body, which is either naked or covered with granules, spines, or scales; in this is contained all the viscera, and from the mouth proceed five long very flexible, simple, or ramified arms, sustained by a series of internal vertebra-like pieces, covered by a naked integument or having granules, scales, or spines, developed from the lateral or inferior parts thereof. The arms of the Ophiuroidea are widely different from the rays of true Star-fishes, which are simple prolongations of the body of the animal, whereas the arms of the Ophiuroidea are superadded to the body, and there is no excavation in them for any prolongation of the digestive organs. There are no ambulacral grooves in the floor of the arms, nor any retractile tubular feet, nor pedicellariæ, in this order.

The mouth, situated in the centre of the under surface of the disk, opens directly into the stomach; this is a sac with one aperture, the mucous membrane of which is covered with vibratile cilia.

The ovaries are placed near the arms, and open by orifices situated at the basal surface of the interbrachial spaces near the mouth.

18

The oral opening is surrounded by five re-entrant angles, which correspond to the intervals of the five arms; from the peristome proceed five buccal fissures, in a line with the axis of the arms; their borders are in general covered with a series of papillæ or plates, and they terminate in a cone of calcareous pieces which perform the part of a jaw. Pl. XIII, figs. 1, 3, 4, 5, 6, show various forms of these dental instruments. At the extremity of each fissure a series of osselets, occupying the interior of the arms, take their origin; their under surface is grooved for lodging a vessel, and between their lateral expansions spaces are formed to receive the base of the fleshy tentacules near the disk.

HISTORY.

The older naturalists united the OPHIURIDÆ to the ASTERIADÆ. Linck[1] first described certain species under the common name *Stella marina,* and designated two species, *S. lacertosa,* and *S. longicauda,* to express the resemblance their arms had to the tail of a lizard; another species, provided with long spines proceeding from the lateral parts of the arms, he called *Rosula scolopendroides,* from the resemblance of the rays to the body of a Scolopendra. The OPHIUROIDEA with ramified arms, as the *Asterophydiæ,* were distinguished by Linck from the simple-rayed forms under the name *Astrophyton.* Seba[2] figured many Ophiuræ among the Star-fishes. Pennant[3] in his 'British Zoology,' and O. F. Müller[4] in his 'Zoologia Danica,' have described many in the genus *Asterias;* and it is under this generic name that all the species of *Ophiuridæ* and *Asterophydia* known to Linné[5] were described, in the thirteenth edition of the 'Systema Naturæ,' by Gmelin. Lamarck,[6] in his 'Système des Animaux sans Vertèbres,' 1801, and in his larger work on the same subject, 1816, established the genera *Ophiura* and *Euryale,* taking *Asterias ophiura,* Müll., as the type of the first, and *Astrophyton scutatum,* Linck, for that of the other. These generic distinctions, first made by Lamarck, were preserved by Delle Chiaje,[7] Risso,[8] and De Blainville;[9] it was not until Professor Agassiz's 'Prodrome'[10] appeared that new generic subdivisions of the order were proposed; in this essay the *Ophiuridæ* were thus defined and classified.

The OPHIURIDÆ differ from the Asteriadæ in this, that the central part of the body

---

[1] Linck, 'De Stellis Marinis,' fol., 42 plates. Lypsiæ, 1733.

[2] Seba, 'Locupletissimi Rerum Naturalium, &c.,' 3 vols. folio, pl. x—xv, 1734.

[3] Pennant, 'British Zoology,' vol. iv, p. 60, 1776.

[4] O. F. Müller, 'Zoologia Danica,' 2 vols. folio. Lips., 1779—1784.

[5] Linné, 'Systema Naturæ,' ed. 13, par Gmelin. 1789.

[6] Lamarck, 'Système des Animaux sans Vertèbris,' 1 vol. 8vo. Paris, 1801.

[7] Delle Chiaje, 'Memorie sulla storia e notomia,' &c. Napoli, 1825.

[8] Risso, 'Hist. Naturelle de l'Europe Merid.,' 5 vols. Paris, 1826.

[9] De Blainville, 'Manuel d'Actinologie,' 1 vol. 8vo, plates. Paris, 1834.

[10] Agassiz, 'Mém. Soc. Sc. Nat. Neuchatel,' vol. i, p. 168. Neuchatel, 1835.

forms a disk, distinct and flat, to which are annexed the rays, more or less elongated, simple, or ramified, and without grooves at their under surface.

1. OPHIURA, Lam., Ag. = (Sect. A, De Blainville).—Disk much depressed; rays simple, squamous; spines very short, embracing the lateral plate.

Types.—*Ophiura texturata,* Lam., *O. lacertosa,* Lam.

2. OPHIOCOMA, Ag., = *Ophiura,* De Blainville (Sect. B).—This genus differs from the preceding by the long moveable spines which arm the rays. Pl. XIII, fig. 4.

3. OPHIURELLA, Ag.—Disk indistinct. All the species are fossil.

O. carinata, Ag. (*Ophiura carinata,* Münst.), *Ophiura Egertoni,* Ag. (*Ophiura Egertoni,* Brod.), Pl. XVII, fig. 4.

4. ACROURA, Ag.—This genus approaches much to the Ophiura properly so called, but it differs in this, that the small scales placed on the sides of the rays replace the spines. The rays themselves are very slender. One species fossil.

A. prisca, Ag. (*Ophiura prisca,* Münst.) Pl. XIV, fig. 5.

5. ASPIDURA, Ag.—A star of ten plates covers the superior surface of the disk, whilst the rays, large in proportion, are surrounded with imbricated scales. One species fossil.

A. loricata, Ag. (*Ophiura loricata,* Goldf.) Pl. XIV, fig. 6.

6. TRICASTER, Ag.; EURYALE, Auct.—Rays bifurcate at their extremity.

T. palmifer, Ag. (*Euryale palmifera,* Lamk.)

7. EURYALE, Lamk.; ASTROPHYTON, Linck; GORGONOCEPHALUS, Leach.;—Disk pentagonal; rays branched from the base.

E. verrucosa, Lam., *E. costata,* Lam., *E. muricata,* Lam.

Müller and Troschel,[1] in 1840, published their first memoir on the classification of the Asteriadæ, in which they grouped the living Ophiuræ into five genera:—1, *Ophiolepis;* 2, *Ophiocoma;* 3, *Ophiothrix;* 4, *Ophioderma;* and 5, *Ophionyx;* comprising twenty-nine species. The same year they added three new genera:—6, *Ophiopolis;* 7, *Ophiomyxa;* 8, *Ophiocnemis.* In 1842 the 'System der Asteriden, by the same authors, appeared, in which three new genera—9, *Ophiarachna;* 10, *Ophiacantha;* and 11, *Ophiomastix,* were added. The EURYALIDÆ comprised three genera—*Asteronyx, Trichaster,* and *Astrophyton,* making in all fourteen genera, and comprising eighty-three species of living OPHIUROIDEA, besides fourteen extinct species distributed in seven genera.

In 1841 Professor Edward Forbes, in his ' History of British Star-fishes,' described thirteen species of OPHIURIDÆ as SPINIGRADE ECHINODERMATA, which he distributed in the genera *Ophiura, Ophiocoma, Astrophyton,* and in 1843,[3] in a memoir communicated to the Linnean Society, he proposed several new genera—1st, *Pectinura* which nearly corresponds to the genus *Ophiarachna;* 2, *Amphiura,* which resembles *Ophiolepis;* and 3, *Ophiopsila,* which comes near *Ophiothrix.* He retained the name

---

[1] ' Archiv für Naturgeschichte,' p. 326.

[2] Edward Forbes, ' History of British Star-fishes.' Van Vorst, 1841.

[3] ' Transactions of the Linnean Society,' London, vol. xix, p. 144.

Ophiura to some of the species placed by Müller and Troschel in their genus *Ophiolepis*.

Many new species have been discovered and described by Sars, Lütkin, Grube, and Philipi, and a few new genera to include the same have been proposed by them as *Ophioplus*, Sars, which also includes *Amphiura*, Forbes, *Ophiocten*, and *Ophiactis*, Lütkin, besides *Ophiopholis* and *Ophiopeza*, Peters.

The fossil species belong to existing genera, as *Ophioderma, Ophiolepis, Amphiura, Ophiocoma ;* and to extinct genera, as *Ophiurella, Acroura, Aspidura, Geocoma, Palæocoma, Aplocoma,* and *Protaster.*   (See Plate XIV.)

On the Skeleton and tegumentary Framework of the Ophiuroidea.

The framework of the *Ophiuridæ* consists of a calcareous or internal, and a tegumentary or external, skeleton.   The former is composed of a number of calcareous osselets of the uniformly reticulate structure common to all the Echinodermata.   These pieces are variously arranged in the different genera.   The disk, in this order, forms the entire visceral cavity, no prolongation of the same extending into the arms, as in the *Asteriadæ ;* on the under surface of the disk five pairs of bones occupy the interbrachial spaces, and form a series of arches that afford attachment to the delicate osselets constituting the skeleton of the upper surface; they are likewise covered with large plates carrying conical-pointed processes, with short spines, which project inwards towards the mouth, and perform the part of jaws and teeth.   Several forms of these dental organs are seen in Pl. XIII, figs. 1, 4, 5, 6.

From the interbrachial spaces five long arms proceed; these organs are supported on a series of osselets resembling the bodies of caudal vertebræ in the tails of lizards ; the scaly character of the tegumentary membrane covering the same increases the resemblance between the ray of *Ophiura texturata*, Lamk., and a lizard's tail.   The brachial osselets have on their anterior and posterior surfaces transverse processes or kinds of articular condyles directed in opposite directions, the one vertical, the other horizontal; two of these transverse condyles always cross each other, so that the lateral flexion of the arms is in nowise impeded; these pieces carry likewise discoidal expansions more or less flexed, which form spaces to receive the base of the fleshy tentacles near the disk.

To the internal skeleton likewise appertain the elongated processes found in the visceral cavity at the base of the arms in relation to the reproductive organs, and the many small osselets arranged on the border of each of the five fissures leading from the mouth-opening to the centre of the base of the arms (Pl. XIII, figs. 4, 5, 6).   Belonging to the same class are the many small pieces piled upon each other at the extremity of the interbrachial bones, and forming the five projecting cones or jaws surrounding the mouth, and which perform the office of teeth (Pl. XIII, fig. 6).

The tegumentary or external skeleton exhibits many interesting modifications in the *Ophiuridæ*. It consists of delicate calcareous plates or scales developed on the disk and arms, the various forms of which and their mode of arrangement afford good characters for generic subdivisions, and constitute the basis of the classification of this natural Order.

In the clothing of the disk we find the character of the tegumentary skeleton, exhibiting many modifications; for example, in *Ophioderma* the surface is covered with numerous granules, and the sides of the arms are furnished with delicate papillæ or fine regular spines; the body is smooth, and the arms resemble the body of a small snake (Pl. XIII, fig. 1).

In *Ophiolepis* (Pl. XIII, fig. 2) the upper surface of the disk is covered with naked scales, having some resemblance in their arrangement to those found on the bodies of lizards. In *Ophiura texturata*, Lamk., the orbicular disk is covered above with a number of small unequal-sized plates, arranged in an imbricated manner, without intermediate smaller pieces. The five pairs of large heart-shaped radial plates are placed close together, the mesial suture of each pair being overlapped by a central imbricated series, which occupy a position above the insertion of the arms. A circle of imbricated scales fills the centre of the disk, and from its circumference a column of round and imbricated scales extends through the centre of the interbrachial spaces to the border, the intermediate area being compactly filled with small elliptical-shaped scales, closely pressed together, as in *Ophiolepis ciliata* (Pl. XIII, fig. 3).

Ophiocoma has the disk more or less uniformly granular in the different species, and even the radial plates (Pl. XIII, fig. 4) are covered with minute granulations. The buccal fissures are entirely bordered with hard papillæ, as in *Ophiocoma dentata* (Pl. XIII, fig. 4), where they form only one range between the teeth and buccal papillæ; in other species they present many varieties of structure. The lateral spines of the arms are smooth and very much developed.

Ophiarachna has the disk granular throughout, except on the ovarial plates, which are naked (Pl. XIII, fig. 5). The buccal plates are divided transversely into a small external piece, and into a larger internal piece. The buccal fissures are provided with dental papillæ, and the arms with delicate, conical, unequal spines, which lie close upon the lateral plates (fig. 5).

Ophiomastix has the disk covered with small imbricated scales, and it likewise supports isolated cylindrical spines. The buccal plates are simple, and the buccal fissures provided with hard papillæ, which are grouped above the dentary pile, as in *Ophiomastix annulosa* (Pl. XIII, fig. 6). The arms are very spinous, and the lateral plates above the rows of spines support a series of claviform pieces with denticulated extremities.

Ophiomyxa has the disk pentagonal, soft, flat, and without granules, and a few obscure scales are observed in the naked integument of *Ophiomyxa pentagona* (Pl. XIII, fig. 7). The buccal plates are nearly round, being a little longer than they are wide.

Ophiothrix has the orbicular or pentagonal disk covered with very fine spines, more or less developed; sometimes they are so delicate that the clothing has quite a villous character; the radial plates are divergent and naked, or partially covered near their base with the general clothing of the disk. *Ophiothrix Rammelsbergii* (Pl. XIII, fig. 8) belongs to the group in which the disk is provided with small short cylinders; the sides form a prominence between the arms; the radial plates are divergent and naked, except near the base, where they are provided with small granulations.

Protaster, an extinct genus, has the circular disk covered with squamiform plates; the genital openings are in the angles of junction of the rays beneath, and the arms are formed of alternating ossicula. Plate XIV, fig. 9, exhibits these characters as shown in *Protaster Sedgwickii*, from the Ludlow rocks of the Upper Silurian series.

Amphiura has an orbicular disk, having its upper surface covered with small smooth scales, the six central plates forming a rosette. The arms, simple and scaly, arise from the centre of the disk, and are provided with lateral subcarinated plates carrying simple lanceolate spines. In *Amphiura tenera* (Plate XIV, fig. 10, *a*, *b*) the disc is small and not lobed, the short delicate arms are surrounded with tri-radiate spines; the radial plates are small, the ventral plates pentagonal, and both are naked.

Aspidura, an extinct form from the Muschelkalk of Germany (Pl. XIV, fig. 6, *a*, *b*) has the upper surface of the body covered with fifteen plates; ten of these form the outer and five the inner circle of the small disk; the arms are furnished with four rows of plates of unequal sizes.

In most genera the upper surface of the disk supports at the base of the arms two large calcareous pieces called radial plates, which are sometimes placed close together, as in *Ophiura texturata*, or apart, as in *Ophiolepis annulosa* (Pl. XIII, fig. 2). They are either entirely naked, as in *Ophiolepis* and *Ophiarachna*, or partly clothed with the general covering of the disk, as in *Ophiothrix*. The radial plates are well preserved *in situ* in many fossil species, as *Ophioderma Milleri* and *O. Gaveyi* (Pl. XVI and XVII).

On the under surface of the disk are five interbrachial spaces between the arms (Plate XIII, figs. 1, 3, 4, 5, 6). In each of these are large smooth pieces, called buccal plates (Mundschilde, scuta buccalia) of authors. They have different forms in different genera; in general, they are single, as in *Ophiolepis* (Pl. XIII, fig. 3), and sometimes they consist of an inner larger and an outer smaller portion, as in *Ophiarachna* (Pl. XIII, fig. 5). In one of these five buccal plates is found, when present, the *umbo*, a small depression in the middle of the plate, the homologue, perhaps, of the madreporiform body of Asteroidea.

Each interbrachial space terminates in a triangular-shaped body, and the five form a star-shaped opening, having the oral aperture for its centre and the buccal fissures for its rays. The terminal process is a narrow cone which rises high up within the mouth, and forms a jaw (maxilla), armed with numerous calcareous pieces piled upon each other, which perform the part of teeth. Plate XIII, fig. 1, shows the form of the maxillæ in *Ophioderma longicauda*; fig. 3, in *Ophiolepis ciliata*; fig. 4, in *Ophiocoma dentata*; fig. 5, in *Ophiarachna septemspinosa*. The lateral parts of the triangular body forming the boundary of the oral fissures (fissuræ buccales) are armed with plates, papillæ, or spines of various forms in the different genera; an inspection of the figures already cited will afford a better idea of the structure and disposition of these parts than any description, however detailed. The border of the fissures is either naked, as in *Ophiothrix* and *Ophionyx*, or provided with hard plates or papillæ (papillæ buccales) arranged in a single row, as in all other genera (see Pl. XIII, figs. 1, 3, 4, 5, 6, 7).

The masticating surface of the maxillæ in all the *Ophiuridæ* carries teeth or papillæ. The teeth occupy the entire breadth of the jaw, and form a perpendicular series of dental processes which extend upwards from the peristome into the interior of the disk. In the genera *Ophioderma* (Pl. XIII, fig. 1), *Ophiolepis* (fig. 3), *Ophiomyxa* (fig. 7), the teeth reach to the mouth-papillæ; but in other genera they are limited to the superior part of the maxillæ, the lower portion being furnished with papillæ, as in *Ophiocoma* (fig. 4) and *Ophiarachna* (fig. 5). The teeth consist of calcareous osselets in general, having a smooth border; seldom is the surface serrated as in *Ophiomyxa pentagona* (fig. 7).

Upon each of the interbrachial spaces, on the under surface of the disk, are found either two or four genital openings (fissuræ genitales); when there are only two, they form long slits, which lie close to the arms. In *Ophiura texturata* the inner extremity of the slit touches the buccal plate, and the outer reaches the circumference of the disk, the internal border of the fissure is finely pectinated with small spines, which fringe the aperture. In the genera with four openings, as *Ophioderma* (Pl. XIII, fig. 1), they are disposed in two behind each other, in the same radial line, or they lie close together, side by side, as in *Ophiocnemis*.

The arms, whether simple, as in *Ophiuridæ*, or ramified, as in *Asterophydiæ*, consist of a great number of jointed osselets, which have been already described; they are clothed externally by a series of plates arranged in an imbricated manner throughout the entire ray, and disposed into dorsal, ventral, and two lateral rows. The dorsal and ventral plates resemble each other in form and covering, and the lateral plates support the spines which arm the rays of most *Ophiuridæ*. The spines are longer or shorter in different genera; they are sometimes stout, or slender, with the surface covered by fine lateral processes, as in *Ophiothrix* (fig. 8); sometimes they are small, short, and closely applied to the arm, as in *Ophioderma* (fig. 1); or they are large, strong, and projecting crosswise, as in *Ophiocoma* (fig. 4), and with additional claviform processes superadded, as in *Ophiomastix* (fig. 6). The character of the spines varies in each genus, and forms a valuable aid in the determination of species. They are moved in general by the contraction of the tegumentary membrane which unites them to the lateral plates.

CLASSIFICATION OF THE OPHIUROIDEA.

The Ophiuroidea are all provided with five arms; in the greatest number they are simple, but in a few they are branched; this character enables us to form a binary division of the order into two families, the OPHIURIDÆ and the ASTEROPHYDIÆ. From the number and position of the genital fissures in the interbrachial spaces, we obtain a character for the division of the families into tribes. The structure of the disk, the presence or absence of scales, plates, or other clothing on the same, the structure and development of the rays and the spines which arm its lateral plates, the structure of the mouth and oral fissures with their armature, the development of the maxillæ and their dentary plates, afford collectively good generic characters.

OPHIUROIDEA.

CHARACTERS.	FAMILIES.
A. Two or four genital fissures. *a.* Arms, five, always simple.	OPHIURIDÆ.
B. Ten genital fissures. *b.* Arms, five, simple or ramified.	ASTEROPHYDIÆ.

1st. FAMILY OPHIURIDÆ.

GENERA.

Four genital fissures in each interbrachial space.
- *Ophioderma.*
- *Ophiocnemis.*

Two genital fissures in each interbrachial space.
A. Disk covered with hard plates.
- *Ophiolepis.*
- *Ophiopeza.*
- *Ophionereis.*
- *Ophiura.*
- *Ophiocten.*
- *Amphiura.*
- *Ophiactis.*
- *Ophiostigma.*
- *Pectinura.*
- *Ophiocoma.*
- *Ophiarachna.*
- *Ophidcanctha.*
- *Ophiomastix.*
- *Acroura.*
- *Aspidura.*
- *Aplocoma.*

Two genital fissures in each interbrachial space.
  B. Disk membranous and naked.

$$\left\{\begin{array}{l}\textit{Ophiomyxa.}\\\textit{Ophioblenna.}\\\textit{Ophioscolex.}\\\textit{Ophiopsila.}\\\textit{Ophiothrix.}\\\textit{Ophiurella.}\\\textit{Protaster.}\end{array}\right.$$

2nd. FAMILY ASTEROPHYDIÆ.

CHARACTERS.

A. Arms simple.

$$\left\{\begin{array}{l}\textit{Asteronyx.}\\\textit{Asterochema.}\\\textit{Asteroporpa.}\end{array}\right.$$

B. Arms ramified and divided.

$$\left\{\begin{array}{l}\textit{Trichaster.}\\\textit{Asterophyton.}\end{array}\right.$$

DESCRIPTION OF THE FOSSIL SPECIES.

Genus OPHIODERMA, *Müller and Troschel*, 1842.

The disk covered with small, close-set granulations; arms long, smooth, and slender; the lateral borders provided with short papillæ or spines closely applied to the lateral plates. Four genital slits in each interbrachial space, disposed in pairs behind each other; two are situated behind the buccal plates, from whence they diverge outwards, and two are placed near the border of the disk. The slits always lie behind each other in the same radial line. The buccal fissures are furnished with small strong papillæ. Pl. XIII, fig. 1, is the disk and rays of *Ophioderma longicauda*, Linck, which exhibit very satisfactorily the organic characters of this group.

The genus *Ophioderma* was established by Müller and Troschel for certain species of *Ophiura* which had for their type the *Stella lumbricalis longicauda*, Linck, tab. xi, No. 17 = *Ophiura lacertosa*, Lamarck.

The Ophiodermas are remarkable for possessing a very smooth body and arms, the granulations and spines are so fine and regular that the rays resemble the skin of an Ophidian reptile, hence the specific name given by Lamarck to Linck's type. Of the fourteen living species, one is found in the Mediterranean and the others in the warmer seas of the globe. The fossil species have, at the present time, been found only in the Lias formation.

OPHIODERMA MILLERI, *Phillips, sp.*   Pl. XVI, figs. 2, 3, *a*, *b*, 4.

ASTERIAS SPHÆRULATA, *Young* and *Bird*.	Geological Survey of the Yorkshire Coast, pl. v, fig. 6, 1822.	
OPHIURA MILLERI,	*Phillips*.	Geol. of Yorkshire, pl. xiii, fig. 20, p. 169, 1829.
— MILLERI,	*Charlesworth*.	London Geological Journal, pl. viii, 1847.
OPHIURELLA MILLERI, *Agassiz*.		Mém. Sc. Nat. Neuchatel, p. 192, 1836.
OPHIODERMA MILLERI, *Wright*.		Ann. and Mag. of Nat. Hist., 2nd series, vol. xiii, 1854.
— MILLERI, *Forbes*.		Morris, Cat. British Fossils, 2nd ed., p. 84, 1854.
— MILLERI, *Wright*.		Brit. Association Report for 1856, p. 402, 1857.
PALÆOCOMA MILLERI, *d'Orbigny*.		Prodrome de Paléontologie, tome i, p. 240, 1850.
— MILLERI, *Oppel*.		Der Juraformation, p. 190, 1856.
— MILLERI, *Dujardin et Hupé*.		Hist. Nat. des Zoophytes Échinodermes, p. 294, 1862.

Disk flat, circular, with ten radial plates, nearly equidistant around the margin; a short series of transverse plates extending between each pair towards the centre of the disk; arms long, round, smooth, tapering from the base to the apex; spines very small, and closely applied to the lateral plates.

Dimensions.—Disk, one inch and three tenths in diameter; arms, four and a half inches in length.

Description.—A very imperfect figure of this Ophiura, under the name *Asterias sphærulata*, was first given by Young and Bird in their 'Geological Survey of the Yorkshire Coast.' Professor John Phillips etched a good outline of it in his 'Geology of Yorkshire,' and named it in that work, without, however, giving any description thereof. Mr. Charlesworth, in the 'London Geological Journal,' 1847, published a beautiful drawing of the fine specimen contained in the museum of the Yorkshire Philosophical Society, and this fossil the council of that institution kindly communicated to me for Pl. XVI, fig. 4, of this work. Although this Ophiura is not uncommon in collections of Yorkshire fossils, still it is rare to find good specimens in which the details of its structure are well preserved.

The disk is flat, circular, or subpentagonal; the radial plates, forming pairs, are placed so far apart that the ten plates are nearly equidistant from each other (Pl. XVI, fig. 4). In this specimen the other parts of the disk are not preserved; in another the body is partly entire (fig. 3, *a*), and shows a series of small transverse plates extending between each pair of radial plates towards the centre of the disk. The arms are long, smooth, and nearly cylindrical; they are three and a half times as long as the diameter of the disk, and taper gently from the base to the apex; their upper or dorsal surface is covered with transverse plates, nearly four times as wide as they are deep (fig. 4); the ventral plates are deeper in proportion to their width, as shown in specimen (fig. 2); the lateral plates are large and imbricated, as seen in a ray from Dr. Bowerbank's specimen (fig. 3, *b*), magnified two and a half diameters. In this fossil the free outer margin of the plates is slightly pectinated, and so fine are the spines that they are only visible when magnified several diameters. In a specimen now before me, showing the base, I find the buccal plates are each composed of four pieces; one central element, which is the largest, is situated nearest the mouth, it has a square form rounded at the oral edge; two lateral pieces are placed on each side of the central element, and one triangular plate at the outer part of the centrum, having its base applied to the central plate, and its apex directed outwards towards the interbuccal spaces. The under surface of the disk, seen between this triangular portion of the buccal plates and the border, is clothed with small close-set imbricated scales. The buccal fissures are very narrow, but their structure is not, for anatomical description, sufficiently exposed.

Affinities and Differences.—In its general characters this species resembles *Ophioderma Gaveyi*, Wr. The disk, however, is proportionately smaller, and the rays are stronger and rounder; the dorsal and ventral plates of the arms are likewise longer transversely, and shorter; the lateral plates are less developed than the homologous parts in *Ophioderma Gaveyi*.

Section of the Marlstone at Rockcliff, near Staithes. (See p. 143.)

No.	LITHOLOGY.	THICK-NESS.	ORGANIC REMAINS.
		Feet.	
	JET ROCK.		
	Lower part of the Upper Lias.	50	*Saurians, Fish, Cephalopoda.*
	MARLSTONE BEDS.		
	Total thickness of the series . . . subdivided into—	160	
	MAIN IRONSTONE BANDS.		
1	Connected blocks of hard ironstone, a foot and upwards in thickness, with thin seams of intervening shale.	25	*Terebratulà trilineata,* Y. and B. = *T. punctata,* Sow.; very abundant.
2	Sandy shale.	5	
3	Iron dogger.		
4	Sandy shale.	10	
5	Iron dogger.		
6	Sandy shale.	15	*Ammonites Clevelandicus,* Y. and B. = *A. margaritatus,* Montf.
7	Iron dogger.		
8	Sandy shale.	18	
9	Iron dogger.		
10	Shaley sandstone.	10	*Am. margaritatus* = *A. vittatus,* Y. and B.
11	Alternations of calcareous sandstone and sandy shale, generally one sandstone bed alternating with a similar bed; the seams covered with fossils.	40	*Belemnites paxillosus,* Schloth., *Pholadomya obliquata,* Phil., *Modiola scalprum,* Sow., *Lima Hermanni,* Voltz., *Ophioderma Milleri,* Phil., *Ophioderma carinata,* Wr., *Ammonites capricornus,* Sch. = *A. maculatus,* Y., *Avicula inæquivalvis,* Sow., *Pecten æquivalvis,* Sow.
12	Shaley marlstone.		*Ammonites maculatus, Cardium proximum,* Hunt.
13	Shaley sandstone, gradually partaking of the nature of Lower Lias shale.	20	*Avicula cygnipes,* Phil., *Cardium truncatum,* Phil., *Rhynchonella tetrahedra,* Sow., *R. acuta,* Sow.
	LOWER LIAS SHALE.		
	Thickness above the level of the sea . . .	150	

Locality and Stratigraphical Position.—*Ophioderma Milleri*, Phil., is found in the same stratigraphical horizon in Yorkshire as *Ophioderma Gaveyi* occupies in the Middle Lias of Gloucestershire. The presence of *Ammonites capricornus*, Schloth. (*Am. maculatus*, Young and Bird), and *Avicula inæquivalvis*, Sow., in the rock containing the large specimen belonging to the museum of the Yorkshire Philosophical Society (Pl. XVI, fig. 4), determines the zone of life to which this Ophiura belongs. It is found in bands of shelly calcareous sandstone belonging to the main Ironstone series or lower portion of the Marlstone beds at Staithes, near Whitby, on the Yorkshire coast, with *Cardium truncatum*, Phil., *Avicula inæquivalvis*, Sow., *Modiola scalprum*, Sow., *Lima Hermanni*, Voltz, and *Pholadomya ambigua*, Sow. (See p. 142.)

The foregoing section, by Mr. L. Hunton,[1] of the Marlstone at Rockcliff, near Whitby, Easington Heights, Mudges Survey, clearly defines the position of the Ophiura-beds in Yorkshire.

History.—This Sea-star was first rudely drawn by Young and Bird, 1822, in their 'Geological Survey of the Yorkshire Coast.' 1829, Professor John Phillips named the species, and gave a good figure of it in his 'Geology of Yorkshire,' but did not describe it. 1836, M. Agassiz, in his 'Prodrome d'une Monographie des Radiares ou Échinodermes,' p. 193, proposed for this and other fossil species the genus *Ophiurella*, without defining the characters on which it was based. 1850, M. A. d'Orbigny, in his 'Prodrome de Paléontologie,' t. i, p. 240, proposed the genus *Palæocoma*, giving *P. Milleri* as the type of the same ; this genus he thus described :—" Ophiures à quatre rangées de pieces aux bras, sans petites pièces intermédiares." 1852, in my 'Contributions to the Palæontology of Gloucestershire,' p. 42, I referred this species to Müller and Troschel's genus *Ophioderma*, associating it with other congeneric forms from the Middle Lias, as *O. Gaveyi, O. Egertoni, O. tenuibrachiata.* In 1854, Professor E. Forbes, in his additions to Morris's Catalogue of British Fossils, and 1857, M. Pictet, in his 'Traité de Paléontologie,' t. iv, p. 274, referred this species to the genus *Ophioderma*. 1862, Dujardin and Hupé, in their 'Histoire naturelle des Échinodermes,' adopted the genus *Palæocoma*, and recorded the species as *Palæocoma Milleri*.

OPHIODERMA EGERTONI, *Broderip*, sp. Pl. XV, fig. 4, *a, b ;* fig. 5.

OPHIURA EGERTONI,	*Broderip.*	Trans. Geol. Soc., 2 series, vol. v, pl. xii, figs. 5, 6, 6*, 1835.
— —	*Forbes.*	Proceed. Geol. Soc., vol. iv, p. 233, fig. 4, 1843.
OPHIODERMA EGERTONI,	*Forbes.*	Morris, Cat. Brit. Fossils, 2nd ed., p. 84, 1854.
— —	*Oppel.*	Die Juraformation, p. 267, 1856.
— —	*Wright.*	Brit. Association Report for 1856, p. 403, 1857.
— —	*Pictet.*	Traité de Paléontologie, tome iv, p. 274, 1857.
— —	*Dujardin et Hupé,*	Hist. Nat. des Èchinod., p. 234, 1862.

[1] 'Trans. of the Geol. Soc. London,' p. 215, vol. v, second series, 1836.

Disk round, small, flat, subpentagonal, almost circular; arms long, smooth, cylindrical, tapering to a filiform termination; vertebral pieces trilobed above.

Dimensions.—Disk, half an inch in diameter; arms, three inches in length.

Description.—This is the most common of our fossil Ophiuridæ, and many fine specimens are obtained from the surface of the large blocks of micaceous sandstone which have fallen from the Star-fish beds of the Middle Lias at Down Cliffs, between Charmouth and Bridport Harbour, on the coast of Dorset. This species resembles *Ophiura texturata*, Lamk., of our present seas, but differs in several important characters from that form.

Some weathered specimens exhibit the structure of the rays in a most beautiful manner; others are so closely invested with the matrix, or covered with a ferruginous crust, that it is impossible to make out their organic details.

The disk is round, and slightly flattened at the interbrachial spaces; its dorsal surface is flat and smooth; the radial plates are large, the pairs closely united by a median suture, and to the adjoining plates by lateral sutures, so that the upper surface of the body appears to be formed entirely by them. The ventral surface exhibits a buccal opening with five rays (fig. 4, *b*); from the base of each arm two long narrow osselets project inwards towards the mouth (fig. 4, *b*). The arms, six times as long as the diameter of the disk, are slender, cylindrical, and taper gradually to the apex; they have on their dorsal surface a series of transverse scales, which, at the base of the arms, are nearly as broad as they are long, in the middle twice as long as they are broad, and proportionately more so towards the extremity; the lateral scales closely embrace the arm, and the spines, if present at all, must have been very short, as I have failed to observe any traces of lateral spines in several well-preserved specimens. The basal scales are nearly the counterpart of the dorsals as to form and structure. (Pl. XV, fig. 5.)

Affinities and Differences.—This species in its general form must have approached very near *Ophiura texturata;* it differed, however, in the size of the radial plates and structure of the disk. It closely resembles *Ophioderma tenuibrachiata* from the same bed, "but in this species the rays are much longer in proportion, and less tapering; they have a more flexible aspect than those of *O. Egertoni*, and present in their section a different form of the central ossicula; for these, instead of being trilobate, are oblong, with a triangular central anterior lobe."—Forbes, 'Proc. Geo. Soc.,' vol. iv.

Ophioderma Egertoni differs from *O. Gaveyi* in having a smaller disk, with much more slender arms, and still more from *O. Milleri*, which is proportionately more robust than *O. Gaveyi.*

Section of the Middle and Upper Lias at Down Cliffs, near Bridport Harbour, Dorset.
(See p. 146.)

	No.	LITHOLOGY.	THICK-NESS.	ORGANIC REMAINS.
			Feet.	
UPPER LIAS, 230 FEET.		Drifted gravel beds.		
	1	Brown sands, sometimes micaceous, with inconstant layers of large sandstone nodules; the fossils are in the uppermost Cephalopoda-bed.	140	*Ammonites opalinus*, Rein., *A. Jurensis*, Zeit., *A. variabilis*, d'Orb., *Nautilus latidorsatus*, d'Orb., *Belemnites, Pecten barbatus*, Sow., *Rhynchonella cynocephala*, Rich.
	2	Dark-greyish clay and sandy marl.	70	
	3	Brownish marly limestone; many of the blocks on the shore are very full of fossils.	2	*Ammonites serpentinus*, Rein., *A. bifrons*, Brug., *A. Raquinianus*, d'Orb., *A. communis*, Sow., *Rhynchonella Bouchardii*, Dav., *Leptæna liassina*, Bouch.
	4	Dark-greyish sandy marl and clay.	15	*Ammonites margaritatus*, Mont.
MIDDLE LIAS, ABOUT 250.	5	Indurated sand, forming large sandstone blocks.	8	
	6	Light-brown sands, more or less indurated, and very micaceous.	60	
	7	Blue marl, forming a well-defined band in the section.	6—8	
	8	Greyish, sandy, laminated marls, with many fossils.	1	*Am. margaritatus*, Mont., *Am. fimbriatus*, Sow., *Pecten æquivalvis*, Sow., *Terebratula cornuta*, Sow.
	9	Foxy coloured, marly sands, with 12-16 inconstant bands of sandy nodules in tiers in the bed.	40	
	10	Thin inconstant band of Crinoidal limestone.	1 in.	*Pentacrinus Johnstonii*, Aust.
	11	Grey, sandy clay, passing into irregular laminated sandstone.	36	*Pecten æquivalvis*, Sow., *Gryphæa Maccullochii*, Sow., fragments of *Pentacrinus*.
	12	The Star-fish bed, hard, grey, micaceous sandstone, falling in large blocks. On the under side of some the Ophiuræ are exposed.	4—6	*Am. margaritatus*, Mont., *A. fimbriatus*, Sow., *Ophioderma Egertoni*, Brod., *Ophioderma tenuibrachiata*, Forbes.
	13	Grey micaceous marls; in the upper part are several rows of fossiliferous nodules.	90?	*Am. margaritatus, A. Normanianus*, d'Orb., *Belemnites clavatus*, Blain., *Pleurotomaria, Chemnitzia*, &c. &c.
		Base of the cliffs.		

OPHIODERMA TENUIBRACHIATA, *Forbes.*   Pl. XVIII, fig. 5, *a, b, c.*

OPHIODERMA TENUIBRACHIATA, *Forbes.*   Proc. Geol. Soc., vol. iv, p. 233, fig. 5.   Read
                                    Nov., 1843.
          —   —   *Charlesworth.*   Lond. Geol. Journ., pl. xix, fig. 1, 1847.
          —   —   *Morris.*   Catalogue of British Fossils, p. 84, 1854.
          —   —   *Pictet.*   Paléontologie, tom. iv, p. 274, 1857.
          —   —   *Wright.*   Brit. Association Report for 1856, p. 403, 1857.
          —   —   *Dujardin et Hupé.*   Hist. Nat. des Échinod., p. 234,
                                    1862.

Disk small, flat, subpentagonal; arms long, delicate, and tapering little.

Dimensions.—Disk, four tenths of an inch in diameter; arms, two inches and seven
tenths in length.

Description.—The body of this species is smaller than that of *Ophioderma Egertoni,*
and the rays in proportion are longer and less tapering. "They have a more flexible
aspect than those of *O. Egertoni,* and present in their section a different form of the
central ossicula; for these, instead of being trilobate, are oblong, with a triangular
central anterior lobe."[1]   In the fine specimen I have figured the ventral surface of the
rays has the marginal angle much more acute than the homologous part exhibits in
O. Egertoni.

Affinities and Differences.—This species may be readily mistaken for *O. Egertoni;*
when carefully examined, however, it is found to possess a smaller disk, longer and less
tapering rays, having a more acute angle at the margins of their ventral surface; a
section of the central ossicula exhibiting a bilobed outline, whereas in *O. Egertoni* a
similar section shows a trilobed form.

Locality and Stratigraphical Position.—This species, like the preceding, is found on
the surface of the large blocks of hard micaceous sandstone described by me as the Star-
fish-bed of the Middle Lias at Down Cliff, near Bridport Harbour, Dorsetshire. (See p.
145.) It is associated with *Ammonites fimbriatus,* Sow., *Amm. margaritatus,* Mont., *Belemnites elongatus,* Mill., and *Ophioderma Egertoni,* Brod.

History.—First discovered by my old friend the late Dr. Murray, of Scarborough, and
communicated by Dr. Bowerbank, F.R.S., to the late Professor Forbes, who first described
it in the 'Proceedings of the Geological Society' for 1843. The very fine specimen
figured in Pl. XVIII belongs to my cabinet.

----

[1] Forbes, 'Proc. Geol. Soc.,' vol. iv, p. 233.

OPHIODERMA GAVEYI, *Wright*, 1852; Pl. XV, fig. 1, *a, b, c, d,* 2, 3; Pl. XVII, fig. 1, *a, b.*

OPHIODERMA GAVEYI, *Wright.* Annals and Mag. of Nat. Hist. 2nd series, vol. xiii,
p. 183, pl. xiii, fig. 1, 1854.
— — *Forbes,* in Morris's Catalogue of Brit. Fossils, 2nd ed.,
p. viii, 1854.
— — *Wright.* Brit. Association Report for 1856, p. 402, 1857.

Disk large, flat, circular; radial plates large, the pairs closely approximated, and separated from the adjoining radial plates by an interval; arms long, slender, tapering gradually; dorsal and ventral plates narrow, the dorsal with a central carina; lateral plates support short, stiff, pectinated spines; base wide, oral opening small, surrounded by five pairs of very prominent tooth-like processes.

Dimensions.—Diameter of the disk, one inch and four tenths; diameter of the rays at their junction with the body, one fourth of an inch; length of the rays, four times the diameter of the disk.

Description.—This Ophiura is found as yet only in one horizon of the Middle Lias, and many fine specimens have been discovered in the original locality near Chipping Camden; the one showing the upper surface (Pl. XV) is in my cabinet, that showing the base belongs to the Museum of the Worcestershire Natural History Society, and I am indebted to my friend Sir Charles Hastings, the President, for his kindness in allowing this fine specimen to be figured in my work.

The disk is large, flat, and circular, slightly inclining to a pentagonal form; it is composed of ten thin, triangular, radial plates, arranged in pairs, each pair forming a heart-shaped shield, with an elevated central carina, formed by the rudimental dorsal plates of the rays covering the prominent vertebral osselets; each pair appear to have been firmly articulated together along the median line, and free from the adjoining pairs at their margins; their surface is smooth, and at the apex of the plates ten elevated eminences indicate the bifurcated terminations of the radial carinæ (Pl. XV, fig. 2). The rays are long, slender, and taper gently; the dorsal plates are small and hexagonal; they are nearly all absent, leaving the osselets thereby exposed (Pl. XV, fig. 1 *b,* 1, *c*). This defect gives a peculiar character to the dorsal surface of the ray, which might be mistaken for a kind of ornamentation; the lateral plates are rounded and closely imbricated, and their free border is toothed with five or six pectinated processes, which in the living state supported spines (Pl. XV, fig. 1, *c,* fig. 2, and fig. 3); the remains of these are sometimes seen attached to their supports; the lateral plates clasp the rays firmly and securely, and overlap the dorsal and ventral plates, the latter are well developed and much elongated transversely; their position and character is well seen in Pl. XV, fig. 1, *a, b, c*.

The buccal plates are absent, but the ten osselets by which the rays were articulated with the disk are well preserved in the specimen figured in Pl. XVII, fig. 1, *a*. These form a considerable star-shaped oral aperture, from the angles of which a tooth-like process projects inwards towards the mouth.

Affinities and Differences.—This Ophiura in its general contour resembles *Ophioderma Milleri*; it is distinguished from that species by having a larger disk, with more slender, tapering rays; the arm-plates are likewise much smaller, and the radial plates of the disk larger. The magnitude of the disk, the structure of the radial plates, and the size of the rays, distinguish it from *Ophioderma Egertoni*.

Locality and Stratigraphical Position.—This Sea-star was discovered by my friend Mr. Gavey, F.G.S., in the Middle Lias of Mickleton Tunnel, near Chipping Campden, Gloucestershire, whilst making the West Midland Railway; it came from the zone of *Ammonites capricornus*; with that Ammonite were associated *Cidaris Edwardsii*, Wr., *Hemipedina Bowerbankii*, Wr., *Uraster Gaveyi*, Forb., *Tropidaster pectinatus*, Forb., and *Pentacrinus robustus*, Wr. Besides these Radiata about sixty species of Mollusca were discovered in the same bed.

I must refer to Mr. Gavey's memoir and section of the cuttings for further details.[1] Fragments of the rays have been found at Hewletts Hill, near Cheltenham, in the same zone of the Middle Lias.

History.—First figured and described in the 'Annals and Magazine of Natural History' for 1852, from specimens kindly sent me by Mr. Gavey. I am not aware that it has been found in any other localities.

OPHIODERMA CARINATA, *Wright*, n. sp.   Pl. XVI, fig. 1, *a, b*.

Disk small, flat, pentagonal; radial plates small, the pairs closely approximated, and separated from the adjoining radial plates by a smooth membranous space; arms long, slender, and gradually tapering; dorsal plates narrow, with an elevated central carina; membranous covering of the disk smooth, and extending like a web between the base of the rays.

Dimensions.—Diameter of the disk seven tenths of an inch; length of the rays, three inches, or about four and a half times the disk's diameter; breadth of the ray at its base one fifth of an inch.

---

[1] " Railway Cuttings at Mickleton Tunnel and Aston Magna," by G. E. Gavey, Esq., F.G.S.; 'Quart. Journ. Geol. Soc.,' vol. ix, p. 29, 1853.

Description.—This **Ophiura** resembles *Ophioderma Gäveyi*, Wr., in the general structure of the disk and rays; the disk, however, is proportionately smaller. The radial plates are closely approximated, and between each pair there is a smooth depressed space; the margin of the disk appears to have been membranous, and extended like a web between the base of the rays (fig. 1, *a*). In the centre of the disk the ten osselets, arranged in pairs, forming part of the buccal framework, are seen projecting upwards (fig. 1, *a*).

The long and slender rays taper gently to their apex; the dorsal plates are narrow, and form a well-marked carina on the middle of the rays. The lateral plates are large, rounded at their free margin, and closely imbricated; there are some obscure indications of small dentations on their outer border for the support of spines (fig. 1, *b*). None of the ventral plates are exposed in the only specimen of this species I have seen.

Affinities and Differences.—This species, in its general form, proportions, and structure, resembles *Ophioderma Gaveyi*, Wr.; it is distinguished from that form chiefly in having a much narrower disk and a stronger carina on the dorsal surface of the rays. The plates of the rays are so imperfectly preserved that their characters cannot be accurately determined, and therefore a comparison with those of *O. Gaveyi* is impossible. It is always prudent to write cautiously about supposed new forms of which we have only a solitary example to examine, for a series of specimens, were they forthcoming, might show that characters supposed to be specific were only varietal; it is with much hesitation, therefore, that I have separated this Ophiura from the preceding on such feeble characters as a narrower disk, and more largely carinated rays.

Locality and Stratigraphical Position.—This Ophiura was collected from the grey micaceous sandstone of the Marlstone at Staithes, where it is very rare; it appears to be the form which was figured by Young and Bird in plate v, fig. 5, in their work on the 'Geological Survey of the Yorkshire Coast,' of which they say—"This is a handsome Star-fish, having five long and bending arms, not unlike some of a smaller size found recently on our shores. It particularly resembles *Asterias sphærulata*, or rather we may venture to pronounce it the same, as it shows the five small beads encircling the mouth. Two of this species are in the Whitby Museum, both in clay-ironstone, occurring in the Alum Shale."[1]

The specimen I have figured belongs to the cabinet of my kind friend, John Leckenby, Esq., F.G.S., who has liberally contributed it with his other rare fossils to this work.

[1] 'Geological Survey of the Yorkshire Coast,' p. 210.

Genus—Ophiolepis, *Müller and Troschel*, 1842.

The upper surface of the disk provided with naked scales or shields (Pl. XIII, fig. 2); two genital slits on each side of the interbrachial spaces close to the arms; oral fissures bordered with a single row of hard papillæ; maxillæ armed with simple prominent dental processes (fig. 2); the lateral scutæ of the arms support papillæ or spines; one or two scales on each tentacule pore; the oral plates are simple and heart-shaped.

The genus *Ophiolepis* forms a natural group, composed of many species, which have been arranged by Müller and Troschel into three sections. The first includes those in which the radial or dorsal plates on the disk are surrounded with a circle of smaller scales; the second in which the small scales are absent; and third those which, besides scales, have rows of spines on the dorsal surface.

> * *Species in which the dorsal plates of the disk are surrounded with small scales.* Type, O. annulosa, Pl. XIII, fig. 2.

> ** *Species in which the dorsal plates are not surrounded by scales.* Type, O. ciliata, Pl. XIII, fig. 3.

> *** *Species in which the disk, besides scales, supports rows of spines.* Type, O. scolopendrica, Linck, tab. xl, fig. 71, 72.

The third section corresponds to the genus *Ophiopholis*, M. and T., into which the species are now merged.

A. *Species of the Lower Lias.*

Ophiolepis Ramsayi, *Wright*, n. sp.　　Pl. XIV, fig. 3, *a, b.*

Disk small; rays short, robust cylindrical; surface of all the scuta covered with fine granulations; the lateral scuta armed with short, stout, thornlike spines.

Dimensions.—Length of the rays eleven twentieths of an inch.

Description.—This beautiful little Brittle-star is sometimes found on the surface of slabs of Lower Lias, associated with portions of the stem of *Pentacrinus tuberculatus,* Mill. The disk appears to have been small; the arms are short, stout, cylindrical, and clothed with a firm armour of prominent plates; examined with an inch-object-glass, their surface is seen to be covered with fine granulations; the dorsal and ventral plates are small, rhomboidal, and much enveloped by the large lateral plates, these carry on each side three or four stiff thornlike spines (Pl. XIV, fig. 3, *a, b*); the robust character of the ray, and the disposition of the spines, is well shown in these figures.

Affinities and Differences.—This species, in the general structure of the arms, resembles *Ophiolepis Murravii*, Forb; but it differs from that form in having them rounder, less tapering, and more moniliform, in consequence of the thickness of the scutal plates; the short stiff spines of the lateral plates I have not seen on *O. Murravii*.

Locality and Statigraphical Position.—I have found this species on the surface of slabs of Lower Lias limestone from Purton passage, near Berkeley, Gloucestershire, associated with *Pentacrinus tuberculatus*, Miller, and young forms of *Ammonites angulatus*, Schloth., and a small smooth *Pecten*, n. sp. My friend, the Rev. P. B. Brodie, F.G.S., collected it from the same horizon at Down Hatherley in the Vale of Gloucester. To his kindness I am indebted for the loan of the specimen figured in Plate XIV, fig. 3, *a, b*.

B. *Species from the Middle Lias.*

OPHIOLEPIS MURRAVII, *Forbes*, sp.   Pl. XIV, fig. 1, *a, b,* fig. 2 ; Pl. XVII, fig. 2, *a, b,* 3, 4. Pl. XIX, fig. 3.

OPHIURA MURRAVII,	*Forbes.*	Proc. Geol. Soc., vol. iv, p. 233, fig. 1. Read Nov., 1843.
—	*Charlesworth.*	Lond. Geol. Journ., pl. xx, figs. 4, 5, 1847.
—	*Morris.*	Catalogue of British Fossils, 2nd ed., p. 84, 1854.
—	*Wright.*	Brit. Association Report for 1856, p. 403, 1857.
—	*Pictet.*	Paléontologie, 2nd ed., tome iv, p. 274, 1857.
OPHIOLEPIS MURRAVII,	*Dujardin et Hupé.*	Hist. Nat. Échinod., p. 245, 1862.

Disk large in proportion to the arms; dorsal surface covered with large scales; radial plates small, scutiform, and projecting on the disk; the converging ossicula at their bases are comparatively large and broad; arms relatively short and tapering. Inferior ray-plates small and triangular. Lateral plates encroaching on those below, and uniting with them beneath in the median line of the ray. They appear to have supported large spines.

Dimensions.—Diameter of the disk seven twentieths of an inch; length of the rays eight tenths of an inch.

Description.—This beautiful little Ophiura is moderately large in proportion to the length of the arms; its upper surface is covered with a series of small imbricated scales, which are disposed in consecutive order from the margin to the centre; a double series of small scales are seen on the interbrachial space of the under surface (Pl. XVII, fig. 2); the arms are short, broad, and tapering, not quite twice and a half the diameter of the disk; the dorsal ray-plates are narrow and heart-shaped, from the manner the lateral plates clasp the rays (Pl. XIV, fig. 1, *b*); the lateral ray-plates are proportionately large, and encroach much upon the ventral series; they have an inflated appearance and appear to have supported numerous small spines, of which some obscure traces only remain; the ventral ray-plates are very small and triangular (Pl. XVII, fig. 2, *b*, fig. 3, and fig. 4); they resemble a series of heart-shaped pieces inserted between the lateral plates, and united with them along the median line. The mouth-opening forms a star with five branches, presenting well-marked buccal fissures (fig. 2, *b*); in one of the specimens it is surrounded with small pieces disposed in a series. In the enlarged figure of this species, given in Pl. XIX, fig. 3, the disposition of the dorsal plates is much better shown.

Affinities and Differences.—The structure of the rays, in this Brittle-star, resembles *Ophiolepis Ramsayii*, Wr., from the Lower Lias; it has similar large lateral and small ventral plates; the absence of the disk, however, in *O. Ramsayii* prevents a more perfect diagnosis of that species, and a more accurate comparison with *O. Murravii*, being made.

Locality and Stratigraphical Position.—This very rare little Brittle-star was discovered by my friend the late Dr. Murray, of Scarborough, to whose liberality I am indebted for the specimens figured in Pl. XIV, fig. 1, and Pl. XVII, fig. 2, both collected from the Marlstone near Staithes, on the Yorkshire coast. The one figured in Pl. XIX was obtained from the Grey Limestone near Scarborough, and kindly communicated by my friend J. Leckenby, Esq., F.G.S., to whose cabinet it belongs.

History.—First figured and described by the late Professor E. Forbes in the fourth volume of the 'Proceedings of the Geological Society,' afterwards by Mr. Charlesworth in the 'London Geological Journal.'

Genus—ACROURA, *Agassiz*, 1834.

This genus, of which the organic characters are very imperfectly defined, was established by M. Agassiz in his Prodrome for *Ophiura prisca*, Münst. (Pl. II, fig. 5), a solitary species from the Muschelkalk. The form approaches much that of *Ophiura texturata*, Lamk., but differs in this, that the rays are long and very delicate; the osselets of the arms are twice as long as they are broad, having their lateral borders incurved (fig. 5 *b*) ; the lateral scuta support a number of small scales instead of spines, as in other genera.

ACROURA BRODIEI, *Wright*, n. sp. Pl. XVII, fig. 5, *a, b, c.*

Disk very small, indistinct ; rays long, very delicate, slightly tapering, nearly a uniform thickness throughout, six times the length of the diameter of the disc ; lateral scutæ with scale-like appendages.

Dimensions.—Diameter of the disk one eighth of an inch ; length of the arms eight tenths of an inch.

Description.—It is with considerable doubt that I have referred to the genus *Acroura* this small *Ophiura*, discovered by my friend the Rev. P. B. Brodie, F.G.S., in the Middle Lias near Cheltenham. Its place there I consider only as provisional until the discovery of better specimens enables us to understand its structure. The disk is extremely small and indistinct ; the rays are long, very delicate, and taper slightly, being nearly of a uniform thickness throughout ; the osselets are nearly twice as long as they are broad, and the dorsal plates are deeper than wide, the few lateral plates that are preserved show small scale-like appendages (fig. 5 *c*). The length of the osselets, and the small scales on the lateral scutes, have induced me to place it in the genus *Acroura.* Under the microscope, armed with an inch-object-glass, the plates appear to be covered with fine transverse lines of ornamentation.

Affinities and Differences.—The smallness of the disk, and the proportionate length of arms, six times as long as the diameter of the body, readily distinguish this species from *Ophioderma tenuibrachiata*, the only species of the Middle Lias for which it could be mistaken ; the extreme delicacy of the rays, and their nearly uniform diameter throughout, with the length of the osselets, form good characters for its diagnosis.

Locality and Statigraphical Position.—This Ophiura was collected by my friend the Rev. P. B. Brodie, from the Middle Lias of Hewletts Hill, near Cheltenham, in the zone of *Ammonites capricornus*, during the excavation of that rock for one of the reservoirs of the Water-company of that town ; it was associated with the stem and side arms of *Pentacrinus robustus*, Wr. I dedicate this species to my friend the Rev. P. B. Brodie, F.G.S., who has kindly communicated for this work the only specimen found.

Genus—OPHIURELLA, *Agassiz.*

Disk small, membranous, and often very indistinct; rays long and slender; lateral ray-plates provided with elongated filiform spines. All the species hitherto discovered belong to the Jurassic series.

OPHIURELLA GRIESBACHII, *Wright.*   Pl. XVIII, fig. 3, *a, b.*

> OPHIODERMA GRIESBACHII, *Wright.*   Ann. and Mag. Nat. Hist., 2nd series, vol. xiii,
> pl. xiii, fig. 2, 1854.
> —   —   *Forbes,* in Morris's Catalogue of Brit. Fossils, 2nd edition,
> additional species, 1854.
> —   —   *Wright.*   Brit. Association Report for 1856, p. 403, 1857.

Disk small, membranous, irregularly pentagonal; rays long, round, slender, and gently tapering; ventral ray-plates moderately large and pentagonal, lateral large, in the form of oblique shields clasping the sides of the rays in an imbricated manner, and supporting short stout spines; buccal opening star-shaped, surrounded by a series of blunt osselets.

Dimensions.—Diameter of the disk, seven twentieths of an inch; length of the rays from peristome to apex, three quarters of an inch.

Description.—This beautiful little Brittle-star of the Oolitic sea was discovered by my friend the late Rev. A. W. Griesbach, of Wollaston, to whose kindness and liberality I am indebted for the series of specimens I possess, and by which I have been enabled to make out the structure of this fossil. The disk is small and often very indistinct, consisting of five pairs of heart-shaped plates, so closely united together that in some specimens it appears to be formed of a single circular disk. Each pair of plates has a heart-shaped form, and the small corresponding ray stånds out in bold relief from the under side of the disk. All the specimens I have seen lie on their upper surface, with the ventral exposed, so that the clothing of the disk is concealed; in one specimen, however, where a portion of one of the plates is weathered, I observed with an inch-object-glass under my microscope the impression of a series of small imbricated scales resting on the matrix. The rays are long, round, slender, and gently tapering, about three times the length of the diameter of the disk; their under surface, the only one exposed, exhibits in weathered specimens a central element of an elongated form, resembling in miniature the elongated centrum of the vertebra of a fish. Pl. XVIII, fig. 3, *b* shows this structure magnified three and a half times. In the only unweathered specimen I possess, from the

Forest Marble, near Wollaston, I find the lateral plates pass round and apparently grasp the sides of the arms in a very distinct hook-like fashion. From the free border of these lateral plates several short thorn-like spines project obliquely outwards; between the central and lateral plates two rows of apertures are formed by the mode these elements of the rays are articulated together; the arms are round, from three to four times the length of the diameter of the body, and taper very gently from the disk to their termination. Since my plates were drawn I have seen a slab, collected from the Forest Marble, near Weymouth, by my late friend the Rev. S. Cooke, F.G.S., on which two specimens of this species are exposed lying on their under side and exposing the upper surface of the disk and arms. The disk is composed of ten radial-plates, united in pairs, and forming five heart-shaped shields, which cover the arms; between them the disk is depressed and undulated at the circumference. On its upper surface several thin small scales are arranged in an imbricated manner and form a radiate pattern towards the centre. The upper plates of the arms consist of two halves united by suture in the middle line; with the lateral plates they curve obliquely outwards clasping round the rays, and terminate beneath the ventral plate, as shown in fig. 3 *b* and fig. 4.

In fig. 3 *a* the under surface of the disk and arms of this species is shown of the natural size. Fig. 3 *b* is the same specimen, magnified three and a half diameters, and in fig. 4 I have shown a portion of one of the arms with the spines articulated to the free margin of the plates, and magnified six diameters.

Affinities and Differences.—This Brittle-star resembles *Ophiolepis Murravii*. It is altogether a more slender form, with a proportionately wider disk having larger and more awl-shaped arms. In the structure of the ray-plates there is likewise a very important difference, *O. Murravii* having the encircling plates on the lateral parts of the rays, with a total absence of the chain-like structure which distinguishes *O. Griesbachii*.

Locality and Stratigraphical Position.—This Ophiurella was first discovered by the late Rev. W. Griesbach in the beds of Forest Marble, near Wollaston, who generously supplied me with several good specimens, one of which I have figured with details.

I have a fair specimen from the Great Oolite of the Windrush Quarries, near Sherborne, Gloucestershire, in a rock similar to that which yielded *Solaster Moretoni*, Forb., so that both these extinct Sea-stars lived together in the same ocean.

My late friend, Rev. S. Cooke, F.G.S., collected several thin slabs of Forest Marble, near Weymouth, on which many beautiful specimens of this species lie in bold relief, and these specimens have enabled me to make out the structure of the upper surface of the disk and rays: all the other specimens which I had previously examined displayed only the ventral aspect of this Ophiurella; which appears to be limited to the upper portion of the Lower Oolitic Formation, as no evidence of its existence has been found in the *Murchisonæ, Sowerbyi,* and *Parkinsoni* zones, which faithfully represent the three divisions of the Inferior Oolite in the English, French, and German Jurassic strata.

The fact of its limitation to the upper portion of the Lower Oolites is another illustration of a former remark that a great break appears to have taken place in the forms of life which prevailed in the upper, when compared with those which existed in the lower portion of the Inferior Oolitic series.

OPHIURELLA NEREIDA, *Wright*, n. sp.　Woodcut figs. 39, 40.

Disk small, irregularly pentalobed, each lobe consisting of a shield-like elevation

FIG. 39.　　　　　　　　FIG. 40.

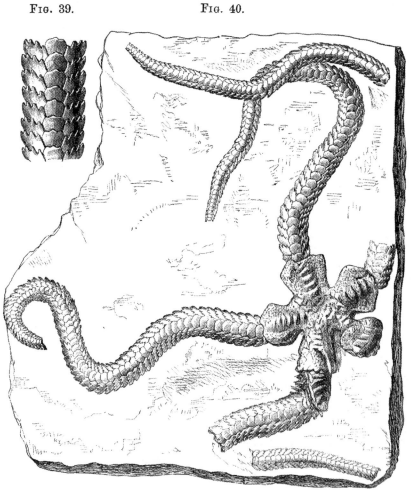

OPHIURELLA NEREIDA, *Wright*.

formed by the radial plates, which are covered by a tegumentary membrane closely studded over with small granules; the interlobular integument is entirely absent, having apparently, if it ever existed, been destroyed in the process of fossilisation.

The arms, five in number, are long, four times the length of the diameter of the disk. They do not taper much between the disk and their termination, and consist of innumerable highly moveable rings, composed of—1st, a centro-dorsal plate, which with its fellows form a long, smooth, convex, continuous chain, flattened at summit, and laid along the middle of the rays; 2nd, of lateral plates which bend downwards, clasping closely the sides of the arms; each of these lateral plates carries a small tubercle, on which stout thorn-like spines are articulated by a kind of ball-and-socket joint; 3rd, the ventral plates, which close in the ray below, are very much concealed; and carry many short stout spines. One of the spiniferous arms of this *Ophiurella*, as it lies on the slab of Calcareous Grit before me, resembles a marine worm, the *Nereis nuntia*, and hence the origin of the specific name I have ventured to give this new Brittle-star. The arms are very much bent and curled, so that this species may be said to have had highly moveable arms.

Dimensions.—Diameter of the disk six tenths of an inch; length of an arm two inches and six tenths of an inch. This is less than they were in the living state, as none of the arms are preserved up to their terminations.

Affinities and Differences.—The fragmentary condition of the disk prevents any definite conclusions as to the true generic position of this form, but it agrees with *Ophiurella* closer than any other. It has the small disk with the upper and under surfaces covered with fine granules; the arms long, compressed, and flattened, the lateral and ventral plates carrying spines, which are specially jointed to the lateral pieces. In all these essential generic characters it agrees with *Ophiurella*. I know of no figured species from the Corallian rocks that resembles our Brittle-star. The only form that occurs to my mind is *Ophiurella bispinosa*, d'Orbig., which has only been named, but was neither described nor figured by the author. Our species is so widely different from all the other described forms that there can be no confusion with them.

Locality and Stratigraphical Position.—This Brittle-star was obtained from the Calciferous Grit at Sandsfoot Castle, Weymouth, by Professor Buckman, F.G.S., who kindly sent it to me for a descriptive note of the species to be inserted in the ' Proceedings ' of the Dorset Natural History and Antiquarian Field Club. The figure was drawn twice the natural size by Mr. Gawan, and has been engraved by Mr. H. P. Woodward. The spines on the lateral plates are not so well shown in the figures as I could have wished, as these spines form a very conspicuous feature in some parts of the rays. Fig. 40 represents *Ophiurella nereida* as it lies on the slab, magnified twice the natural size; and Fig. 39 is a portion of one of the rays still more highly magnified to show the arrangement and form of the dorsal and lateral plates of the arms. The specimen belongs to Professor Buckman's collection.

Genus—Amphiura, *Forbes*, 1844.

Discal body orbicular, having its upper surface covered with small, smooth plates, the six central plates forming a rosette. The simple scaly rays arising from the centre of the disk are provided with lateral subcarinated plates, carrying from 5—12 simple lanceolate spines. Ambulacral papillæ either absent, single, or double. In *Amphiura tenera* the disk is small and not lobed, the short delicate arms are surrounded with triradiate spines; the radial plates are small, the ventral plates pentagonal, and both are naked.

Amphiura Prattii, *Forbes.*   Pl. XVIII, figs. 1 *a—d*, 2.

Amphiura Prattii, *Forbes.*   Proceed. Geol. Soc., vol. iv, p. 233, 1844.

Disk with a few smooth imbricated scales on the intermediate plates beneath. Arms in small specimen nearly six times as long as the diameter of the disk; slender and flexible, and not tapering much. The inferior ray-scales are quadrangular, with oblique sides. Each lateral ray-plate bears a row of slender, conical, diverging, smooth spines, which are about as long as the breadth of the ray, and there is also a very small spine at the inferior angle of each.

Dimensions.—Fig. 1. Diameter of the disk unknown; length of the rays from the peristome to the apex one inch and three tenths. Fig. 2. Disk small, rays six times as long as the diameter of the body.

Description.—This rare Ophiurid presents only the under surface of the animal, the disk, which was probably soft or cartilaginous, having disappeared, though its limits are well marked by the forms of the ossicula composing the bases of the arms which were inserted into it. There are traces of a few smooth imbricated scales on the intermediate plates beneath. The smaller specimen coiled up on a slab (fig. 2, natural size) shows how small the discal body was in proportion to the length of the arms, which were at least six times as long as the diameter of the disk. The under side of the arms is composed of a central plate of hexagonal shape, with an ambulacral hole on each side (fig. 1 *a, b, c, d*). The lateral plates are imbricated and their free border is provided with five or six stiff thorn-like spines (fig. 1 *b, c, d*), in which the structure of the arm is satisfactorily shown. The foramina between the central and lateral plates are well seen in these figures. The small specimen (fig. 2) does not exhibit much of its structure, and we are at present ignorant of the anatomy of the upper surface of the disk and rays. I have not seen any other example of this rare form than the one I have figured, which belongs to the British Museum. Fig. 1 *a* shows the under surface of the natural size ; fig. 1 *b, c,* and *d,* are portions of the rays magnified four times to exhibit details of structure. Fig. 2 is a small specimen of the same species coiled up on a slab.

Stratigraphical position.—This curious Brittle-star was discovered by the late Mr. Pratt, F.G.S., in the Oxford Clay, and presents most of the characters of the genus *Amphiura* (' Linnean Transactions,' vol. xix, part ii, p. 150), to which it was referred by the late Professor Edward Forbes, F.R.S., in his notice of the species, in the ' Proceedings of the Geological Society,' vol. iv, p. 233.

ADDENDA.

Species of the Genus OPHIOLEPIS.

A. *Species from the Inferior Oolite.*

OPHIOLEPIS LECKENBYI, *Wright*, n. sp.   Pl. XIX, fig. 3 *a, b.*

Disk small, covered with ten pear-shaped radial scales, arranged in pairs at the base of the rays. The converging apices of the scales appear to have left a small space in the centre of the disk, the integument of which is absent. Arms moderately long, and stoutly built of rings of imbricated scales, arranged in a chain-shaped style of grouping; the lower surface of the arm-scales have a supplementary piece inserted at the heart-shaped angle formed at the junction of the lateral scales of each arm along the mesial line. Arms tapering very gently towards the termination.

Dimensions.—Diameter of the disk four tenths of an inch; length of the rays one inch and three tenths.

Description.—This beautiful little Brittle-star was collected by the late Peter Cullen from the Grey Limestone near Scarborough. It was kindly communicated to me by my old much-esteemed friend the late lamented John Leckenby, Esq., F.G.S., to whose memory I dedicate the species. As the Leckenby collection now forms part of the Woodwardian Museum, Cambridge, the type-specimens will be found there. The example before me is rather smaller than the type figured in Pl. XIX, fig 3 *a, b,* of the natural size.

The small disk was covered above with ten radial, pear-shaped scales, grouped in pairs, and placed over the base of the rays. They must have covered the greater portion of the disk; unfortunately the central part is absent, and the only other specimen which I possess lies upon its upper surface, and displays only the under side.

The rays were flat, moderately long, flexed, and tapering very gently from their discal to their terminal end. They were stoutly built of rings of imbricated scales, which are neatly arranged so as to form a chain-like structure (fig. 3 *a*), each upper segment being made up of two halves united along the middle line, as shown in fig. 3 *b*. The under segment is formed likewise of two pear-shaped, obliquely-twisted scales, which at their junction in the mesial line have a small supplementary scale inserted between the round terminations of the lateral pieces. The rays appear to have been much flexed, as all the arms are bent (fig. 3 *a*) in different directions in the smaller specimen on a slab of Grey Limestone now before me.

Affinities and Differences.—*Ophiolepis Leckenbyi*, Wr., differs much from *Ophiolepis Murravii*, Wr. The disk is smaller, and the arms longer and more slender, and thus this species is easily distinguished from the Lias forms.

Locality and Stratigraphical Position.—The two specimens of this Brittle-star were collected from the Grey Limestone near Scarborough, which represents the middle zone (*Stephanoceras Humphriesianum*) of the Inferior Oolite. These little radiates appear to be rare, as I have seen no other examples of them in any other local collection.

B. *Species from the Upper Triassic.*

OPHIOLEPIS DAMESII. *Wright*, n. sp. Pl. XXI, figs. 4 and 5.

> OPHIOLEPIS DAMESII, *Wright.* Zeitschrift der Deutschen geologischen Gesellschaft, pl. xxix, figs. 5 *a*, *b*, p. 821, Jahrgang 1874.

Body-disk small, upper surface convex and undulated, rays long, cylindrical, four times the length of the diameter of the disk; dorsal shields of rays smooth, semicircular; ventral shields form a double chain of round prominent joints, which extend from the mouth-opening to the end of the rays.

Dimensions.—Diameter of the disk one quarter of an inch. Length of the rays one inch.

Description.—This beautiful little Brittle-Star belongs to the genus *Ophiolepis*. The upper surface of the discal body is convex and undulated, being elevated at the points in the circumference where the rays proceed outwards, and depressed and concave between the inter-radial elevations. The round slender rays end in very fine points. The upper surface of the discal body is covered with small delicate scales. The rays have semi-circular shields on their upper surface, and on the under side a double row of articulated segments, which extend from the mouth to the end of the rays.

This *Ophiolepis* resembles *Ophiolepis Murravii*, Forbes, from the Marlstone of Yorkshire, but differs from it by the greater proportional length of the rays and more slender structure of the same; from *O. Ramsayi*, Wr., by the absence of the thorny processes which project from the free angles of the lateral shields.

Stratigraphical Position.—This Brittle-star was first discovered by Professor H. Roemer between the upper and lower Bone-bed breccias, near Hildesheim, with other fossils belonging to the *Avicula contorta* beds. Several of these specimens were sent to me for examination and description by Dr. Dames, of Berlin, and my notes on the same were afterwards published, with good figures, in the 'Zeitschrift der Deutschen geologischen Gesellschaft' for 1874.

A few months after I returned the Ophiolepis to Dr. Dames at Berlin, several specimens were submitted to me which had been collected from the *Avicula contorta* beds at Garden Cliff, on the Severn, from the dark shales above the Bone-bed. I had no difficulty in making out the identity of this species with the one I had so recently described from Hildesheim, and I brought the facts before the Cotteswold Naturalists' Club at one of the regular meetings of the Society, anent the discovery of this little Brittle-star in beds of the same age, and at points so far remote from each other as Westbury on the Severn, and Hildesheim in Northern Germany. Very soon after this I was informed that the same Ophiolepis had been collected in the black shales above the Bone-bed, near Leicester. These three discoveries of this Brittle Sea-star, in beds which had never previously yielded any remains of Echinoderms, made quite an epoch in the history of the *Avicula contorta* series, inasmuch as a doubt had been entertained as to the nature of the conditions under which the *Avicula contorta* series had been deposited: many of the fossils were small dwarfed individuals which it was conjectured might have lived under lacustrine conditions. The discovery of a true marine radiate in these shales afforded, therefore, positive proof of the conditions under which the Bone-bed breccia and its overlying shales had been formed, and recalled to my mind the important observation my late colleague, Professor Edward Forbes, F.R.S., had made on first seeing the *Avicula contorta* and White Lias beds at Lyme Regis: writing to Professor Ramsay, F.R.S., in 1847, he said:

" My visit to Lyme gave me a thoroughly clear idea of the Lias, and the succession of its fossils, which I much wanted. I now can picture to myself all the events of its formation. At the base of it I saw the so-called White Lias, which, so far as I have seen, seems to me to be essentially different from the Lias, and possibly the terminating strata of the Triassic series. I broached a notion to Sir Henry De la Beche, from what I saw, that the red marls were formed in a great salt inland sea (a sort of Aralo-Caspian), during the last state of which the White Lias was formed, that the bed was then either elevated and converted into land, or depressed and turned into a part of the ocean, when the Liassic fauna came in. This notion is not merely hypothetical; the fossils of the White Lias (very few in species) suggested the idea; they are curiously representative of the existing Caspian fauna. Such a state of things would account for the general and hitherto almost unaccountable unfossiliferous character of the Trias in our area, and for the extinction of the last traces of the palæozoic fauna."[1]

In connection with the subject of the British fossil forms of the Ophiuridæ described in this Monograph, attention may be drawn to the peculiarities of structure presented by the anatomy of the living *Astrophiura permira*. Mr. W. Percy Sladen,[2] F.L.S., states that with regard to the latter.

1. The combination of Ophiuroid disk- and arm-structure within a pentagonal Asteroid form of body.

2. The character of the ambulacral system is Asteroid: the divisional plates not only being homologous with, but resembling in the manner of their disposition, the ambulacral plates of Asteroidea; at the same time furnishing a highly suggestive representation of their phylogenetic development.

3. The rudimentary structure of the mouth-armature is more Asteroid than Ophiuroid in general facies. Absence of teeth, jaw-plates, and jaws.

4. An extension of the peritoneal cavity to the extremity of the functional portion of the rays, that is to say, to the margin of the pentagonal body.

5. The condition and aborted character of that portion of the brachial series which is prolonged beyond the body-disk is extremely rudimentary.

6. The continuity of the tentacular pore-system is limited to the disk only.

The above characters, as Mr. Sladen points out, are clearly sufficient to stamp the peculiarity of this extraordinary Echinoderm; and, whilst excluding it from any known group of genera by their remarkable nature and by the aberrant departure they present from all previous types, are such as would seem to necessitate the relegation of the form to a family apart by itself.

To speak definitely, he adds, as to the exact position of intermediacy which the organism holds between the Asteroidea and Ophiuroidea would obviously be premature, without a more detailed examination of the internal anatomy than the present specimen in its dry condition will permit, as well as some knowledge of the life-history of the form. It may, however, be safely affirmed, without overstepping the bounds of due caution, that *Astrophiuria* bridges further over, from the Ophiuroid side, the differences which have separated the two orders, than any previously described Starfish or Brittle-star.

[1] 'Memoir of Edward Forbes, F.R.S.,' by Dr. Wilson and Prof. Geikie, p. 418.

[2] " On *Astrophiura permira*, an Echinoderm-form intermediate between Ophiuroidea and Asteroidea," by W. Percy Sladen, F.L.S., F.G.S. Received June 18th, 1878. Communicated by Professor Duncan, F.R.S. 'Proc. Roy. Soc.,' 20th June, 1878.

ADDITIONAL GENERA OF THE FAMILY GONIASTERIDÆ.

Genus—STELLASTER, *Gray*, 1841.

Body nearly pentagonal, flat on both sides, with two wide rows of large granulated marginal plates, both of which contribute to the formation of the high border. Each ventral marginal plate carries a suspended spine. Both surfaces of the disk are covered with granulated plates; vent subcentral. The ambulacra are narrow and the suckers biserial.

The genus *Goniaster* was proposed by Professor Agassiz, in his 'Prodrome,' to include Starfishes with a large pentagonal body, having the margin bordered by a series of wide thick plates or tesseræ, superimposed in pairs, and which support spines, granules, &c. The upper and under surface of the disk is covered with small polygonal plates, set closely together like a mosaic, and fitted into this marginal framework; the surface of the ossicles is smooth or covered with granulations. The ambulacral furrows are narrow, with two rows of pedal suckers therein; the vent opens near the dorsal surface, and the mouth-opening is slit-like and pentagonal.

Müller and Troschel, in their 'System der Asteriden,' suppressed the genus *Goniaster*, and instead thereof erected the genus *Astrogonium*, adding also their new genus *Goniodiscus* and Gray's *Stellaster*.

The diagnostic characters of these groups were chiefly obtained from the structure of the marginal plates and their appendages, the figure and arrangement of the discal plates, and the form and development of the rays, and may be thus described:

1. *Astrogonium*.—Marginal plates large and smooth towards the centre, their inner border encircled by granules.

2. *Goniodiscus*.—Marginal plates having the entire upper surface covered with close-set granulations.

3. *Stellaster*.—Marginal plates granulated, the ventral segment supporting a pendant spine, the rays elongated and tapering to a lanceolate extremity.

Many of these characters are absent in fossil *Goniasteridæ*, and are, therefore, value-less for palæontological purposes; for this reason I have retained—

(A) The genus *Goniaster* for the large pentagonal short-rayed forms, and

(B) The genus *Stellaster* for those with a smaller disk and more elongated rays.

This division must be considered merely provisional until we become better acquainted with the comparative anatomy of extinct forms.

The very fine fossil discovered by my friend Samuel Sharp, Esq., F.G.S., in the

Ironstone beds of the Inferior Oolite near Northampton, appears to belong to the group *Stellaster*, in consequence of the smallness of the disk and the length and development of the rays. The absence of pendant spines or any indication of their presence warns us, however, to be cautious in drawing hasty conclusions as to the true generic position of this Starfish, seeing that the presence of this kind of spine is considered to be diagnostic of living Stellasters. Whether this fossil ever possessed such a spine or not the mould does not enable me to make any positive statement about.

STELLASTER SHARPII, *Wright,* n. sp. Pl. XX.

Body pentagonal, sides arched, rays much elongated and tapering to a narrow extremity; marginal plates thick, surface of the same finely granulated. Under surface of the disk covered with small, close-set, polygonal ossicles, having had apparently a very granular surface. The circumference of each ray surrounded by sixty pairs of marginal plates, which extend from the centre of the arch of one interradial space to the same point of the adjoining area. Ambulacral channels narrow, oral opening large.

Dimensions.—Diameter of the disk two inches from the centre of one areal arch to the same point on the opposite one; from ray point to ray point six inches, depth of the border at the centre of the arch three tenths of an inch.

Description.—This remarkable fossil is entirely a mould in Ironstone, none of the ossicles having been preserved; but the sharp impressions of their forms and sculpture impressed on the Ironstone reveal a tolerably correct idea of the anatomy of the plates.

The Starfish rests upon its upper surface, which is firmly imbedded in the matrix, so that the size, shape, and character of the dorsal ossicles still remain to be discovered.

Those on the under side of the disk are nearly uniform in size, and are small, pentagonal, and hexagonal. The granulations on the surface of these small bones appear to have been very large, whilst those on the marginal tesseræ were very small.

Affinities and Differences.—This species resembles *Stellaster Berthandi,* from the ' Calcaire à Entroques,' Mâcon; it differs, however, from that form in being larger, and in having the interradial spaces more arched and the rays themselves much larger. But the marginal tesseræ and discal ossicles are very much alike.

The mould, however, does not give the character of the granulations. It differs from *Goniaster obtusus,* Wr., from the Inferior Oolite of Crickley, in having longer and more lanceolate rays, and from *Goniaster Hamptonensis,* Wr., from the Great Oolite of Minchinhampton, in the greater width of its tesseræ, and length of its rays.

STELLASTER BERTHANDI, *Wright*, n. sp.   Pl. XXI, fig. 2.

Body pentagonal, sides with arches much flattened; tesseræ thick and narrow, thirty-six to forty pairs around the margin of one ray; under surface of the disk covered with small, close-set, equal-sized ossicles; ambulacral channels wide; dorsal discal ossicles absent.

Observation.—Since Mr. Sharp's specimen came into my hands for description a plaster-mould has been kindly communicated to me by Professor Berthand, of Mâcon, Saone-et-Loire, France. The original was collected from the 'Calcaire à Entroques,' Mâcon. I mention it here, in connection with *Stellaster Sharpii*, Wr., as showing that *Goniaster* was a type of the Asteriadæ which prevailed during the first stage of the Jurassic period, as the three forms we now know are all specifically distinct, and belong to the lower division of the Oolitic series.

Genus—URASTER, *Agassiz*, 1835.

URASTER SPINIGER, *Wright*, n. sp.   Pl. XXI, fig. 1.   Woodcut, fig. 41.

Rays five, short, broad, curved, and petaloidal; ambulacral areas wide; margins bordered by a series of small ossicles, which form beaded ridges on each side of the ambulacral spaces. Ossicles support numerous small, short, blunt spines, which lie in profusion on the sides of the rays, and similar spines appear to have clothed the dorsal surface, and are seen " *in sitû* " in the twisted portion of one of the rays, as depicted in the annexed figure.

The disk small in proportion to the width of the rays and diameter of the Starfish.

Dimensions.—Diameter of the disk one half inch; length of each ray one inch; breadth of a ray at its widest part four tenths of an inch.

Description.—This Starfish was obtained from the Forest Marble, near Road, Wilts, where it was collected by Dr. H. F. Parsons. The specimen came into my hands in a very bad state, but by carefully backing it up with plaster of Paris I have been able to develop a considerable portion of its anatomy. The small ossicles which occupied the central portion of the ambulacral areas are absent, and there remains only the vacant spaces they at one time filled. The margins forming the boundaries of the areas are built up of a series of small ossicles, which are largest at the discal end, and diminish in size towards the termination of the rays. They form a beaded structure of considerable strength, which supported likewise a great number of small, stout, blunt spines, that appear to have passed round the sides of the rays and clothed the dorsal surface, as a few are seen *in situ* on the integument. This spinous condition of the tegumentary membrane has suggested the specific name given to this new form of Uraster.

Affinities and Differences.—This Starfish differs so much from the other fossil species of the genus *Uraster* that it may possibly prove to be the type of a new genus, when more details are learned anent the anatomy of the skeleton by the discovery of new materials. In the mean time I have grouped it with the Urasters which it so closely resembles in its general characters, whilst it differs in others, as, for instance, the beaded ridge of ossicles which bound the ambulacral area.

Locality and Stratigraphical Position.—*Uraster spiniger* was collected from the shelly beds of the Forest Marble, near Road, Wilts. It appears to be rare, as I can find no record of any other Star-fish found in the fossiliferous beds of that locality.

The annexed beautiful figure was drawn on wood by the late Mr. C. R. Bone, and engraved by Mr. Lee, for the 'Transactions of the Cotteswold Naturalists Field Club,'

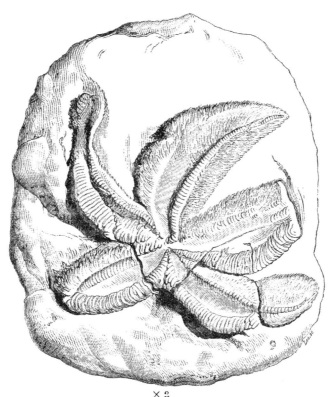

× 2

Fig. 41.—Uraster spiniger, *Wright*. Magnified two diameters.

to illustrate my notes on the specimen which I read at one of the meetings of the Club some time ago. I am indebted to my old kind friend, Dr. W. H. Paine, the worthy

Honorary Secretary of the Club, for permission to use the woodcut to illustrate my description of the species in this Monograph, for which I return Dr. Paine my very best thanks. The figure is drawn to scale twice the natural size and is reversed.

ADDITIONAL NOTES ON GENERA BELONGING TO THE FAMILY ASTROPECTINIDÆ, *Müller and Troschel*.

ASTROPECTEN RECTUS.   Pl. XIX, fig. 1 *a, b.*

The specimen of this species, figured in Pl. XII, is only a horizontal section of the skeleton just as it was entombed in the Calcareous Grit. Since that plate was drawn I found in the collection of my late friend Mr. John Leckenby, F.G.S., a fragment which shows the structure of the marginal ossicles, and is now figured in Pl. XIX, so that we are able to make a tolerably perfect restoration of this fine Star-fish of the Corallian seas. Each of the marginal plates appears to have carried a kind of socket, probably for the articulation of a spine. They all have rounded margins and a cancellated structure, so well depicted by my late able and accurate artist friend, Mr. C. R. Bone, in fig 1 *a* and fig. 1 *b.*

Fig. 1, Plate XII, was drawn from a specimen in the collection of the Rev. T. Wiltshire, F.G.S.

ASTROPECTEN HOOPERI, *Wright*, n. sp.   Pl. XXI, fig. 3 (p. 123).

FOSSIL ASTERIA.   Loudon's Mag. of Nat. Hist., vol. ii, p. 73, fig. 19, 1829.

This beautiful Astropecten was found by the Rev. James Hooper, Rector of Stawell, at Horsington, Dorset, and a good sketch of the original fossil was communicated by W. H. R. N., Yeovil, 21st August, 1828, to Mr. Loudon's 'Magazine of Natural History.' "The Starfish was taken from a stratum of Cornbrash, and is a very perfect specimen. The sketch and the figure are the exact size of the original."

Many years ago, when my 'Monograph on the Fossil Asteroidea' was passing through the press, I made several inquiries about this specimen, which I failed to trace, as the figure at that time and since had been entirely overlooked. I have, therefore, had a copy made of the woodcut in Loudon's Magazine, as it is the only Star-fish I know from the Cornbrash, and it is right that it should find a place among its congeners in this Monograph.

Diagnosis.—Rays five, short, acutely lanceolate; sides straight, intermediate angles obtuse; marginal plates rounded, and diminishing gradually in size from the angle to the apex. Each marginal plate appears to have supported on its upper surface a thorny spine, which projected obliquely backwards.

On Trichotaster, a New Genus of Silurian Asteroidea.

On the 26th March, 1873, Dr. Grindrod, of Malvern, forwarded to me some Silurian fossils to examine and determine for him.

On one of the slabs I discovered a remarkable little Starfish on a fragment of Wenlock Limestone from Dudley, forming a *new genus* of the order Asteroidea.

Genus—Trichotaster, *Wright*, 1873.

Small Starfishes with a large disk, the structure and clothing of which are unknown, this part being absent. Rays numerous, short, lanceolate, and closely surrounding the disk in a stellate fashion, the border of each armed with small triangular dentiform spines; from the extremity of each ray a stem-like multi-articulate process proceeds, equal in length to the ray itself. The outer rings of this process support slender-jointed lateral appendages, imparting a tuft-like character to the terminations of the rays.

Trichotaster plumiformis, *Wright*, n. sp.

Body oblong, rays ten, unequal, the side of each border surrounded with from eight to ten separate, sharp, triangular, dentiform spines.

The terminal stem, consisting of eight joints, of which the four outer or last-formed support two slender, lateral, antennæform processes; the length of these diminish from the proximal to the distal pair, and produce by this arrangement a beautiful plume-like structure at the termination of each ray.

Dimensions.—Disk, greatest diameter four-tenths of an inch. Lesser diameter three tenths of an inch. Length of one ray, with its terminal plume-like process, three-tenths of an inch.

Locality and Stratigraphical Position.—Imbedded in hard grey Wenlock Limestone from Dudley. I have only seen the one specimen now described.

APPENDIX I.

SUMMARY* OF THE BRITISH LIASSIC AND OOLITIC ECHINOIDEA, ASTEROIDEA, AND OPHIUROIDEA,

Described in Volumes I and II of the Oolitic Echinodermata.

Order I.—ECHINOIDEA, *Wright* (Vol. I, pp. 4, 6—16).

Body-shell (*test*) spheroidal, oval, cordate, or depressed, without arms, furnished with a distinct mouth (*oral opening*), whose border (*peristome*) is sometimes simple, sometimes lobed, always on the under side, and generally armed with five calcareous sets of plates (*jaws*). Anal opening variously situated on the upper or under side, or on the marginal border. Body enclosed in a shell (*test*), composed usually of twenty, sometimes of more than twenty (as in the family of the Palæozoic *Perischoechinidæ*), columns of calcareous plates, forming, in either case, ten areas. Five of the areas (*ambulacral*) containing each two rows of apertures (*poriferous zones*) for the passage (in the living state) of retractile suckers (*ambulacral tubes*). The other five areas (*interambulacral*) destitute of sucker-pores. Ambulacral pores disposed in single pairs (*unigeminal*), double (*bigeminal*), or triple oblique (*trigeminal*). Ambulacral pore-columns (*areas*) sometimes continuous from the peristome to the summit (*complete*), sometimes confined to the upper surface of the test (*interrupted*), or forming re-entering curves (*petaloid*). Surface of test studded with tubercles (*primary, secondary,* and *miliary*), possessing spines of various forms and dimensions. Spines articulated on the rounded upper part of a tubercle (*mamelon*), which rises from a conical process (*boss*). Base of tubercle surrounded by a round, oval, smooth, excavated space (*areola* or *scrobicule*). Summit of test marked by an apical disc, composed generally of five genital and five ocular plates, usually in contact and central. Cutaneous surface of shell, especially near the mouth, bearing small tripartite, pincer-like bodies (*pedicellariæ*), placed on a short stalk, whose lower portion encloses a calcareous nucleus. Pedicellariæ capable (in living state) of seizing small bodies and passing them from one to the other. Movement of the animal effected by the motion of the spines and the ambulacral tubes.

The ECHINOIDEA (including the Perischoechinidæ) range from the Silurian to the

* Compiled by the Rev. T. Wiltshire, M.A., F.G.S., Hon. Sec. Pal. Soc.

recent period, and are represented in the British Liassic and Oolitic strata by nine families, twenty-one genera, and 120 species, and are divisible into two (A and B) sections, the ECHINOIDEA ENDOCYCLICA and the ECHINOIDEA EXOCYCLICA.

Section A.—Echinoidea endocyclica, *Wright* (Vol. I, p. 17).

Anal opening within the genital plates, always opposite the mouth. Jaws always present.

The section contains five families: CIDARIDÆ, HEMICIDARIDÆ, DIADEMADÆ, ECHINIDÆ, and SALENIADÆ, ranging from the Trias to the existing period. The five families are represented in the British Liassic and Oolitic strata, and give thirteen genera and eighty-two species.

FAMILY I.—CIDARIDÆ, *Wright* (Vol. I, p. 23).

Test thick, spheroidal, generally depressed at the upper and the under surfaces. Ambulacral areas narrow, usually undulating, and destitute of primary tubercles. Inter-ambulacral areas wide, carrying a few large primary perforated tubercles. Poriferous zones narrow, straight; pores generally unigeminal. Oral and anal openings large. Peristome destitute of notches. Jaws present. Primary spines long, massive, and more or less cylindrical. Family ranging from the Trias to the present period, and represented in the British Liassic and Oolitic strata by three genera: CIDARIS, RABDOCIDARIS, and DIPLOCIDARIS, with fifteen species.

Genus 1.—CIDARIS, *Klein* (Vol. I, p. 25).

Test thick, more or less depressed. Ambulacral areas undulating. Primary tubercles few, rarely more than six in a row. Miliary zones more or less wide. Pores of the poriferous zone unigeminal and contiguous. Range of genus from the Trias to the recent period. Eleven British Liassic and Oolitic species (Vol. I, pp. 26—53, 451).

Genus 2.—RABDOCIDARIS, *Desor* (Vol. I, p. 54).

Test thick, slightly depressed. Ambulacral areas nearly straight. Miliary zones wide. Pores unigeminal, not contiguous, but connected by a small horizontal furrow. Range of genus from the Liassic to the Lower Cretaceous beds. Two British Liassic and Oolitic species (Vol. I, pp. 54, 55).

Genus 3.—DIPLOCIDARIS, *Desor* (Vol. I, p. 56).

Test thick, large, depressed. Pores arranged in double oblique pairs. Genus found in the Liassic and Oolitic strata. Two British Liassic and Oolitic species (Vol. I, pp. 56—58, 452).

FAMILY II.—HEMICIDARIDÆ, *Wright* (Vol. I, p. 68).

Test thick, spheroidal, more or less depressed. Genital plates sometimes bearing perforated tubercles. Ambulacral areas, narrow, straight, or undulating, and carrying perforated semi-tubercles. Interambulacral areas wide, bearing large tubercles, generally perforated and raised on prominent bosses. Poriferous zones narrow and undulated; pores small, contiguous, and unigeminal, except near the peristome, where they are bigeminal and trigeminal. Primary spines generally long, cylindrical, and tapering, sculptured with lines, and generally without asperities. Mouth and anal opening large. Peristome divided by notches into five large and five small lobes. Jaws present. Range of family from the Oolitic to the Tertiary strata. Family represented in the British Oolitic strata by one genus, Hemicidaris. Thirteen British Oolitic species.

Genus 4.—HEMICIDARIS, *Agassiz* (Vol. I, p. 69).

Test thick, generally flattened at base. Ambulacral areas more or less undulating, bordered by minute tubercles, and supporting on the lower part, for about a quarter of the whole length, perforated tubercles, which increase in magnitude from below upwards, and on the other three quarters minute perforated tubercles. Thirteen British Oolitic species (Vol. I, pp. 69—100, 453).

FAMILY III.—DIADEMADÆ, *Wright*, (Vol. I, p. 106).

Test in general moderately thick, subpentagonal, more or less depressed. Ambulacral areas straight, more or less wide, furnished with two or four rows of tubercles, often as large as those of the interambulacral. Interambulacral areas equal to, or double as wide as, the ambulacral. Tubercles of the interambulacral areas either primary, of equal size, in two to eight rows, perforated, or with two to four rows of secondary tubercles. Poriferous zones narrow, and almost always straight; pores unigeminal, bigeminal, or trigeminal. Oral and anal opening large. Apical disc small. Peristome decagonal, generally deeply notched. Spines cylindrical, solid or tubular; those of the fossil not longer than the diameter of the shell; surface of the solid forms covered with very fine, minute longitudinal striæ; and that of the tubular with oblique annulations of fringe-like scales. Jaws present. Range of family from the Trias to the existing period. Family represented in the British Liassic and Oolitic strata by four genera: PSEUDODIADEMA, HEMIPEDINA, HETEROCIDARIS, and PEDINA, in thirty-four species.

Genus 5.—PSEUDODIADEMA, *Desor.* (Vol. I, p. 108).

Test moderately thick, not large. Ambulacral areas one third to one half the width of the interambulacral. Ambulacral areas furnished with two rows of primary tubercles. Interambulacral areas sometimes provided with two rows of primary tubercles, sometimes showing four or six rows of equal-sized tubercles at the equator. Pores of poriferous zones unigeminal throughout, or bigeminal in the upper part of the zones. Apical disc small. Oral opening large. Peristome deeply notched. Spines solid, cylindrical or needle-shaped, short, and covered longitudinally with very minute microscopic lines. Range of genus from the Lias to the Cretaceous beds. Thirteen British Liassic and Oolitic species (Vol. I, pp. 110—132, 456).

Genus 6.—HEMIPEDINA, *Wright* (Vol. I, p. 143).

Test thin, much depressed on the upper surface, and flat or slightly concave on the lower surface. Ambulacral areas narrow. Interambulacral areas usually double the width of the ambulacral, with two to eight rows of perforated tubercles at the equator, having uncrenulated bosses; upper part of miliary zone wide and covered with fine granulations. Pores of poriferous zones unigeminal. Spines slender and needle-shaped. Range of genus from the Lias to the Oolites. Eighteen British Liassic and Oolitic species (Vol. I, pp. 144—167, 457).

Genus 7.—HETEROCIDARIS, *Cotteau* (Vol. I, p. 455).

Test thin, large, circular, depressed on the upper surface, almost flat on the lower surface. Ambulacral areas very narrow, furnished with two rows of small distinct perforated tubercles, uniform in size, and raised on small bosses placed in regular rows. Interambulacral areas very wide, provided with from six to eight rows of large, nearly equal-sized perforated tubercles, raised on prominent bosses with crenulated summits. Miliary zone finely granulated. Poriferous zones narrow; pores small, non-conjugate, having a slight disposition to a trigeminal arrangement near the oral opening. Oral opening large, pentagonal, and lobed. Anal opening circular. Spines long and cylindrical, covered with fine longitudinal lines and small indistinct tubercles. The genus occurs in the Inferior Oolite. One British Oolitic species (Vol. I, p. 456).

Genus 8.—Pedina, *Agassiz* (Vol. I, p. 171).

Test very thin, nearly equally depressed on the upper and under surfaces. Ambulacral areas narrow, with two rows of small primary marginal tubercles, Interambulacral areas wide, with two rows of perforated primary tubercles, and two to four rows of secondary tubercles extending from the peristome to the circumference. Bosses of tubercles uncrenulated. Poriferous zones wide; pores trigeminal, and arranged obliquely. Apical disc small. Oral opening small, slightly notched. The genus is met with in the Oolitic strata. Two British Oolitic species (Vol. I, pp. 173—176, 459).

Family IV.—ECHINIDÆ, *Wright* (Vol. I, p. 183).

Test generally thin, globular or depressed; plates numerous. Ambulacral areas one third the width of the interambulacral, bearing two or more rows of tubercles. Interambulacral areas with large plates, sometimes perforated at the angles, sometimes marked by depressions on the line of sutures, sometimes sculptured with irregular figures in relief, sometimes microscopically plaited, and generally bearing many small imperforate tubercles arranged in rows. Poriferous zones narrow or wide; pores unigeminal, bigeminal, obliquely trigeminal, or in three vertical rows. Spines short, subulate, and sculptured with fine longitudinal lines. Oral opening large or small. Peristome often pentagonal, feebly or strongly notched. Jaws present. Apical disc small. The family ranges from the Oolitic to the existing period. Family represented in the British Oolitic strata by four genera: Glypticus, Magnotia, Polycyphus, and Stromechinus. Ten British Oolitic species.

Genus 9.—Glypticus, *Agassiz* (Vol. I, p. 185).

Test thick, small, spherical, or subconoidal. Ambulacral areas narrow and straight, with two rows of marginal tubercles arranged in linear series. Interambulacral areas bearing two rows of well-developed imperforate and uncrenulated tubercles on the under side, and irregularly formed prominences on the upper surface. Poriferous zones narrow; pores unigeminal. Oral opening wide. Peristome decagonal, slightly notched, and unequally lobed. Genus found in the Oolitic strata. One British Oolitic species (Vol. I, p. 186).

Genus 10.—MAGNOTIA, *Michelin* (Vol. I, p. 190).

Test thin, small, inflated, with a concave base. Ambulacral areas narrow and straight. Interambulacral areas marked by a deep median depression, and bearing many small, equal-sized, imperforate, and uncrenulated tubercles. Poriferous zones narrow; pores unigeminal. Genus found in the Oolitic strata. One British Oolitic species (Vol. I, p. 191).

Genus 11.—POLYCYPHUS, *Agassiz* (Vol. I, p. 196).

Test thin, small, hemispherical. Ambulacral areas narrow and straight. Interambulacral areas wide, with a slight median depression. Both the ambulacral and the interambulacral areas are covered with numerous small equal-sized, regularly-arranged, imperforate tubercles, which are much larger on the under side than on the upper surface. Poriferous zones wide and depressed. Pores in triple oblique pairs, which become most numerous near the peristome. Oral opening wide. Peristome pentagonal, slightly notched, and marked by unequal lobes. Genus found in the Oolitic strata. Two British Oolitic species (Vol. 1, pp. 197—199).

Genus 12.—STOMECHINUS, *Desor* (Vol. I, p. 203).

Test thin, of moderate size, globular, conoidal, more or less depressed. Ambulacral areas narrow, one third the width of the ambulacral, bearing two marginal rows of numerous small tubercles, and sometimes two additional rows of a smaller size. Interambulacral areas wide, bearing two rows of larger tubercles and several rows of smaller tubercles. Miliary zones sometimes broad and granular, sometimes narrow and naked. Poriferous zones moderately wide; pores arranged in triple oblique rows. Mouth opening large. Peristome subpentagonal, deeply notched, and unequally lobed. Spines short, stout, and bluntly pointed. Genus found in the Oolitic strata. Six British Oolitic species (Vol. 1, pp. 204—217).

FAMILY V.—SALENIADÆ, *Wright* (Vol. I, p. 226).

Test thin, small, spheroidal or depressed. Ambulacral areas always narrow, straight, or flexuous, with two rows of small tubercles, which alternate with each other on the margins of the areas. Interambulacral areas wide, with two rows of large primary tubercles, perforated or imperforated, with bosses having crenulated summits. Poriferous zones narrow; pores unigeminal, except near the peristome, where they are obliquely trigeminal. Oral opening variable in size. Peristome more or less decagonal, deeply or feebly notched. Jaws present. Apical disc usually greatly developed, with an additional suranal shield, sometimes formed of one piece, sometimes of from three to eight pieces. Spines long and slender, covered with fine longitudinal striæ. Range of family from the Lias to the existing period. One British Liassic and Oolitic genus, ACROSALENIA, with ten British species.

Genus 13.—ACROSALENIA, *Agassiz* (Vol. I, p. 229).

Test thin, spheroidal or depressed. Ambulacral areas narrow, straight, or slightly undulated, with two rows of small crenulated and perforated tubercles on their margins. Interambulacral tubercles perforated, raised on large prominent bosses, with crenulated summits. Jaws present. Apical disc moderately small and not prominent. Sur-anal shield composed of one or many pieces. Anal opening slightly excentric. Range of the genus from the Lias to Lower Cretaceous. Ten British Liassic and Oolitic species (Vol. I, pp. 230—249, 460—462).

Section B.—**Echinoidea exocyclica,** *Wright* (Vol. I, p. 17).

Anal opening outside the genital plates, never opposite the mouth.

The section contains eight families, Echinoconidæ, Collyritidæ, Echinonidæ, Echinobrissidæ, Echinolampidæ, Clypeasteridæ, Echinocoridæ, Spatangidæ, and ranges from the Lias to the existing period. Four families (ECHINOCONIDÆ, COLLYRITIDÆ, ECHINOBRISSIDÆ, and ECHINOLAMPIDÆ) are represented in the British Liassic and Oolitic strata by eight genera: HOLECTYPUS, PYGASTER, HYBOCLYPUS, GALEROPYGUS, COLLY-RITES, ECHINOBRISSUS, CLYPEUS, PYGURUS; and by thirty-eight species.

FAMILY VI.—ECHINOCONIDÆ, *Wright* (Vol. I, p. 258).

Test thin, circumference circular or subpentagonal. Upper surface conical; under surface somewhat flat. Ambulacral areas narrow, simple, and lanceolate. Interambulacral wide, bearing small perforated tubercles, arranged more or less regularly, and supported on bosses with smooth or crenulated summits. Poriferous zones continuous from oral opening to apical disc; pores unigeminal, except near the mouth, where they are in triple oblique pairs. Oral opening circular, central, or subcentral. Peristome more or less divided by notches into ten lobes. Jaws present. Anal opening large and excentrical, oblong, dorsal, marginal, infra-marginal, or basal, sometimes occupying the entire space from the mouth to the border. Apical disc, mostly central and vertical, with five ovarian and five ocular plates, the right lateral plate large, and supporting a prominent spongy madreporiform body. Spines small, short, and covered with microscopic lines. The Echinoconidæ range from the Lias to the existing period. Four British Oolitic genera (HOLECTYPUS, PYGASTER, HYBOCLYPUS, and GALEROPYGUS), with thirteen British Oolitic species.

Genus 14.—HOLECTYPUS, *Desor* (Vol. I, p. 259).

Test thin, circumference circular, more or less hemispherical, depressed, always tumid at the sides, flat or concave at the base. Ambulacral areas narrow and lanceolate, bearing six to eight rows of small tubercles from base to apex. Interambulacral wide, supporting numerous small perforated tubercles, arranged in vertical and concentric rows. Poriferous zones narrow; pores unigeminal throughout. Oral opening very large, circular, in centre of base. Peristome divided into ten equal lobes. Jaws present. Anal opening very large, inferior, infra-marginal, rarely marginal, sometimes occupying the entire space between the oral opening and the border. Apical disc nearly central and, vertical. The genus is found in the Oolites and Lower Cretaceous beds. Three British Oolitic species (Vol. I, pp. 260—268).

Genus 15.—PYGASTER, *Agassiz* (Vol. I, p. 273).

Test thick, subpentagonal, more or less elevated, convex on upper surface, concave on under. Ambulcaral areas narrow, bearing four to six rows of small tubercles, the marginal rows extending from base to apex. Interambulacral wide, bearing small, nearly equal-sized, perforated, and uncrenulated tubercles, arranged in vertical and horizontal rows, of which the two are median, and continuous throughout. Poriferous zones narrow ; pores unigeminal throughout. Oral opening circular. Peristome deeply notched and marked by ten equal lobes. Jaws present. Anal opening very large, oblong, superior, and almost universally continuous with apical disc. Apical opening central, large. Spines small, short, covered with minute longitudinal lines. Range of genus from the Lias· to the existing period. Five British Oolitic species (Vol. I, pp. 275—282, 463).

Genus 16.—HYBOCLYPUS, *Agassiz* (Vol. I, p. 291).

Test thin, subcircular or subovate, upper surface unequally elevated, the anterior half usually higher than the posterior half. Under surface concave or undulating. Ambulacral areas narrow, flexuous, and disjointed at summit by the length of the apical disc. Three terminate at the oral extremity of the apical disc, two at the opposite end.

Surface of areas covered with numerous concentric rows of small perforated tubercles, set close together on low crenulated bosses. Poriferous zones very narrow ; pores simple and unigeminal, placed close together on the upper surface and wide apart at the base. Oral aperture excentric near the anterior third of the base, oval, with the longer axis oblique. Peristome unnotched, subpentagonal. Jaws absent. Anal aperture elongate, excentric, in a longitudinal valley in the upper surface of the test, and in contact with the apical disc. Apical disc central, but not vertical, elongated. Anterior pairs of plates large, and disposed side by side between the anterior and posterior ovarials. The posterior oculars at the extremity of the posterior ovarials. Genus found in the Oolitic strata, Three British Oolitic species (Vol I, pp. 298—303, but *not* pp. 292—296, *see* p. 465).

Genus 17.—GALEROPYGUS, *Cotteau* (Vol. I, p. 465).

Test thin, similar to Hyboclypus, but possessing a subcompact, instead of an elongated apical disc. Ovarian plates arranged in crescentric form around the concave anterior opening of the round discal aperture, the right antero-lateral plate the largest, supporting the madreporiform body; plates small, presenting acute angles, inserted into the notches of the interambulacral segments of the discal opening. Ocular plates very small, inter-calated between the angles of the ovarian plates. Range of genus from Lias to Oolitic strata. Two British Oolitic species (Vol. I, pp. 292, 296; *see* pp. 465, 466).

FAMILY VII.—COLLYRITIDÆ, *Wright* (Vol. I, p. 304).

Test thin, ovoid, elongated or cordiform, depressed. Ambulacral areas narrow, and so arranged that three meet at the larger portion of the apical disc, and two nearer the anal opening, over which they form an arch. Ambulacra covered with small perforated crenulated tubercles. Poriferous zones narrow; pores unigeminal. Oral opening small, oval, or circular, placed towards the sulcated border. Peristome entire. Jaws absent. Anal opening small and oval. Apical disc elongate, subcentral, divided into two portions. Three of the ocular plates are placed at the junction of the three ambulacra and two at the junction of the other ambulacra. Family found in the Oolitic and Cretaceous strata. One British Oolitic genus (COLLYRITES) and three species.

Genus 18.—COLLYRITES, *Deluc* (Vol. I, p. 307).

Test thin, ovoid, subovoid, or cordiform, depressed, slightly sulcated at the border nearest the oral opening. Ambulacal areas narrow, straight, the five not in contact, at the apex, two meeting at the summit, and three nearer the anal opening. Poriferous zones narrow, continuous ; pores unigeminal. Interambulacral areas covered with small perforated and crenulated tubercles. Oral opening small, obtusely pentagonal, excentric. Jaws absent. Anal opening small, oval in the middle of non-sulcated margin. Genus found in the Oolitic and Cretaceous strata. Three British Oolitic species (Vol. I, 309—318).

FAMILY VIII.—ECHINOBRISSIDÆ, *Wright* (Vol. I, p. 330).

Test thin, oblong, subpentagonal, circular, covered uniformly with small uncrenulated imperforate tubercles, largest on the under side. Ambulacral areas narrow and petaloid, lanceolate above, expanded in the middle, contracted and open below. Poriferous zones at the border and base narrow ; pores small, equal. Interambulacra wide. Apical disc small. Oral opening small, subcentral, and pentagonal, sometimes lobed. Jaws absent. Anal opening large, placed in a furrow on the upper surface or at the margin. The Echinobrissidæ range from the Oolitic to the existing period. Two British Oolitic genera (ECHINOBRISSUS and CLYPEUS), and sixteen species.

Genus 19.—ECHINOBRISSUS, *Breynius* (Vol. I, p. 331).

Test thin, small, oval or subcircular, convex above, subconcave below, rounded in front, truncated, and generally widest behind ; marked by small tubercles, which are larger on the under surface than on the upper. Ambulacral areas petaloid on upper surface, straight on lower. Oral opening small, excentric. Peristome pentagonal or oblique. Jaws absent. Anal opening circular, on upper surface, in a sulcus sometimes extending the whole distance, sometimes only a part, from the apical disc to the border of under side. Apical disc small, square, compact. Genus ranges from the Oolitic to the existing period. Eight British Oolitic species (Vol. I, pp. 332—353).

Genus 20.—CLYPEUS, *Klein* (Vol. I, p. 360).

Test large, discoidal ; upper surface moderately subconvex, depressed ; under surface flat, concave, undulated. Ambulacra petaloid on upper surface, straight on under surface. Oral opening central or subcentral, small, lodged in a depression. Peristome lobed. Jaws absent. Anal opening oval or pyriform, situated in a sulcus extending wholly or partially from the apical disc to the lower border. The genera is found in the Oolitic strata. Eight British Oolitic species (Vol. I, pp. 361—382, 466).

Family IX.—ECHINOLAMPIDÆ, *Wright* (Vol. I, p. 389).

Test thin, oval or subpentagonal; upper surface depressed, convex, or conoidal; vertex usually excentric; surface bearing small, often perforated, tubercles. Ambulacral areas petaloid on upper surface, straight or slightly petaloid on lower. Oral opening small and lobed. Jaws absent. Anal opening marginal, supra-marginal, or infra-marginal, not in a sulcus. The Echinolampidæ range from the Oolitic to the existing period. One British Oolitic genus (PYGURUS) and six species.

Genus 21.—PYGURUS, *d'Orbigny* (Vol. I, p. 391).

Test thin, large, discoidal, subpentagonal; upper surface slightly conical, under surface concave and undulating, rostrated behind, sulcated in front. Tubercles small, perforated, larger on under than upper surface. Ambulacral areas distinctly petaloid on upper, subpetaloid on lower surface. Pores in a single series, until near the oral opening, there becoming closely crowded in triple oblique ranks. Oral opening pentagonal and excentric, strongly lobed. Jaws absent. Anal opening oval, infra-marginal, sometimes corresponding to, sometimes transverse to, the direction of the longest axis of the base of the test. Apical disc small. Genus found in the Oolitic and Cretaceous strata. Six British Oolitic species (Vol. I, pp. 392—405, 467).

Order II.—ASTEROIDEA, *Wright* (Vol. I, p. 4, Vol. II, pp. 1—22).

Body stellate, depressed, provided with five or more hollow arms (*rays*) containing (in the living state) prolongations of the viscera. Mouth (*oral opening*) always central on the under side, and sometimes serving as an anal opening as well. Anal opening, when present, subcentral on the upper side. Skeleton consisting of many solid calcareous pieces, variable as to number, size, and position. Integument (*perisome*) coriaceous (in the living state), and studded with calcareous spines of various forms, and also with tubercles carrying a crown of short bristly spines (*paxillæ*). Rays grooved and pierced at the centre of the under surface with two or four sets of pores for the admission (in the living state) of retractile tubular suckers, which pass between the edges of internal ossicles and not through them. Upper surface of body marked by one or more madreporiform tubercles near the angle between two rays. Eyes (in the living state) generally present at the extremity of the rays. No dental apparatus. Small, pincers-like bodies supported on slender flexible stems (*pedicellariæ*), on the integument surrounding the mouth and bases of spines. Movement of the body performed by the suckers and spines.

The Asteroidea range from the Cambrian formation to the existing period, and are represented in the British Liassic and Oolitic strata by two sections (A and B), four families URASTERIDÆ, TROPIDASTERIDÆ, GONIASTERIDÆ, and ASTROPECTINIDÆ), eight genera (URASTER, TROPIDASTER, SOLASTER, GONIASTER, STELLASTER, LUIDIA, PLUMASTER, and ASTROPECTEN), and twenty-three species.

Section A.—Four rows of pores in each ambulacral space.

The section is represented in the British Liassic and Oolitic strata by one family (URASTERIDÆ) and three species.

FAMILY X.—URASTERIDÆ, *Wright* (Vol. II, p. ii).

Body stellate, rays long, rows of spines near the ambulacral areas on the under side. Upper surface of body and rays covered with blunt or pointed spines, either scattered, grouped, or linearly arranged. Integument between the spines naked. Four rows of pores in each ambulacral space. Pedicellariæ on soft stems. Anal opening subcentral. Range of the family from the Lias to the existing period. Family represented in the British Liassic and Oolitic strata by one genus (URASTER) and three species

*Genus 22.—*URASTER, *Agassiz* (Vol. II, p. 99).

Body stellate, rays five, moderately long, cylindrical or lanceolate, deeply cleft on under side, fringed below with rows of small and laterally with larger spines. Skeleton composed of small irregularly shaped and femur-like ossicula, articulated together in a retiform manner. Upper surface of body studded with blunt or pointed spines, scattered or grouped together in tufts, and arranged more or less regularly in longitudinal rows. Ambulacral avenues wide, composed internally of two rows of long femur-like bones, spaced for the admission of four series of tentacula. Anal opening small, subcentral. Madreporiform body simple. Range of the genus from the Lias to the existing period. Three British Liassic and Oolitic species (Vol. II, pp. 100, 101, 166).

Section B.—Two rows of pores in each ambulacral space.

The section is represented in the British Liassic and Oolitic strata by three families (TROPIDASTERIDÆ, GONIASTERIDÆ, and ASTROPECTINIDÆ), seven genera (TROPIDASTER, SOLASTER, GONIASTER, STELLASTER, LUIDIA, PLUMASTER, and ASTROPECTEN), and twenty species.

Family XI.—TROPIDASTERIDÆ, *Wright* (Vol. II, p. iii).

Body stellate, rays variable in number, short. Upper surface spinous. Ambulacra bordered by transverse plates with spiniferous crests on their anterior margin. Two rows of pores on each ambulacral space. Anal opening present. Family ranging from the Lias to the existing period. Represented in the British Liassic and Oolitic strata by two genera (Tropidaster and Solaster) and two species.

Genus 23.—Tropidaster, *Forbes* (Vol. II, p. 102).

Body stellate, rays five, short, convex, bearing on the upper surface a central carina of a double series of squamose plates with lateral ranges of spines. Ambulacral grooves rather broad, lanceolate. Ambulacral ossicula oblong, rather broad, strongly bent and crenated on the sulcal border. Marginal plates crested and bearing a row of spines. Two rows of ambulacral pores. Anal opening on dorsal surface (?). Genus found in the Lias. One British Liassic species (Vol. II, p. 102).

Genus 24.—Solaster, *Forbes* (Vol. II, p. 104).

Body stellate, rays numerous, short, bearing on the upper surface fasciculated spines. Ambulacral grooves wide and deep. Marginal plates crested, bearing a row of long slender acicular spines. Two rows of ambulacral pores. No pedicellariæ. Anal opening central. Range of genus from the Oolitic to the existing period. One British Oolitic species (Vol. II, p. 104).

Family XII.—GONIASTERIDÆ, *Wright* (Vol. II, p. 7).

Body pentagonal, angles more or less produced, somewhat flattened on both sides, margins conspicuous. Upper and under surface central. Ossicula small, flat, hexagonal, pentagonal, tetragonal, covered with granules, spines, or pedicellariæ. Marginal plates in a single series on each surface, large and granulated. Ambulacral avenues bordered by a series of square ossicula, often marked by parallel grooves for lodging the spines. Two rows of ambulacral pores. Anal opening on upper side, excentric. Family ranging from the Lias to the existing period. Represented in the British Oolitic strata by two genera (Goniaster and Stellaster) and three species.

Genus 25.—Goniaster, *Agassiz* (Vol. II, p. 106).

Body pentagonal, only slightly produced, flattened on both sides.   Margins bounded by two rows of granulated plates, always larger than the remaining ossicula.   Two rows of ambulacral pores.   Range of genus from Lias to existing period.   Two British Oolitic species (Vol II, pp. 108, 109).

Genus 26.—Stellaster, *Gray* (Vol. II, p. 164).

Body pentagonal, angles produced, rays tapering.   Marginal plates thick, finely granulated.   Lower marginal plates bearing spines and granules.   Upper and under surfaces, apart from the marginal plates, covered with small polygonal ossicles, smooth and granulated.   Ambulacral furrows narrow, with two rows of ambulacral pores.   Anal opening subcentral.   Range of genus from the Oolitic to the existing period.   One British Oolitic species (Vol. II, pp. 165, 166).

Family XIII.—ASTROPECTINIDÆ, *Müller* and *Troschel* (Vol. II, p. 110).

Body stellate, flattened on both sides, rays five or numerous, narrow, elongate, and bordered, sometimes by an upper and sometimes by an upper and an under series of large conspicuous plates.   Lower marginal plates always spiniferous; upper, when present, granulated and spiniferous to a greater or less degree.   Upper surface, apart from the marginal plates, covered with closely-set paxillæ; under surface crowded with short spines arranged in regular rows.   Ambulacral furrows narrow.   Two rows of ambulacral pores.   Separate anal opening absent.   Family ranging from the Lias to the existing period.   Represented in the British Liassic and Oolitic strata by three genera (Luidia, Plumaster, and Astropecten) and fifteen species.

Genus 27.— Luidia, *Forbes* (Vol. II, p. 110).

Body stellate, covered on upper surface with paxillæ; disc comparatively small; rays variable in number, sometimes numerous, elongate, provided on under surface with a single row of marginal plates, carrying at the central portion short thick spines, and at the margin long recurved spines.   Ambulacral furrow narrow.   Two rows of ambulacral pores.   Range of genus from the Lias to recent period.   One British Liassic species (Vol. II, p. 111).

Genus 28.—PLUMASTER, *Wright* (Vol. II, p. 112).

Body stellate, rays numerous, elongate, expanded in middle, provided on under side with a single row of marginal plates, transversely elongate, slightly arched, pectinated on one border, and displaying a single row of tubercles carrying long spines. Ambulacral furrow·narrow and depressed. Radial ossicles at the angle of the rays very prominent and sculptured. Two rows of ambulacral pores. Genus found in the Liassic strata. One British Liassic species (Vol. II, p. 112).

Genus 29.—ASTROPECTEN, *Linck* (Vol. II, p. 113).

Body stellate, depressed, rays five, elongate. Margin of rays provided with two rows of marginal plates, those on under side bearing spine-like scales, which increase in size from within outwards, and terminate in long spines; those on upper side covered with granules and often becoming spinous. Upper surface of body and rays (apart from the marginal plates) clothed with granules or plates, which are crowned with groups of minute spines. Ambulacral furrows more or less broad. Two rows of ambulacral pores. Range of genus from the Lias to existing period. Thirteen British Liassic and Oolitic species (Vol. II, pp. 113—129, 168).

Order III.—OPHIUROIDEA, *Wright* (Vol. I, p. 5; Vol. II, p. 131).

Body (*disc*) discoidal, depressed, having long slender arms (*rays*), destitute of an excavation for the viscera. Upper surface of disc supporting at base of arms ten large calcareous pieces (*radial plates*), sometimes close together, sometimes entirely naked, sometimes partly clothed with the general covering of the body. Under surface of disc bearing large smooth pieces (*buccal plates*) of different forms, generally single, and ten in number, sometimes doubled, and consisting of an inner larger and outer smaller portion. Under surface of disc also marked in each of the angular spaces between the arms (*inter-brachial*) by genital openings (*fissures*), either two or four, or ten. Mouth (*oral opening*) always central, on the under side provided with a masticatory apparatus consisting of a cone of calcareous pieces, and always serving for the anal opening. Skeleton of many calcareous pieces variable as to number, size, and arrangement, and covered with an integument, either naked or bearing granules, scales, or spines. Rays simple or ramified, enclosed in four sets of jointed calcareous plates, one above, one below, and one on each side; the lower set single, double, or quadrupal; spines developed on the sides; under surface not grooved and not pierced for tubular feet. Madreporiform tubercle on under side in one of the buccal plates placed near the oral opening. No pedicellariæ. Movement (in the living state) effected by the rays.

The Ophiuroidea range from the Silurian formation to the existing period, and are represented in the British Liassic and Oolitic strata by one family (Ophiuridæ), five genera (Ophioderma, Ophiolepis, Acroura, Amphiura, and Ophiurella), and twelve species; and are divisible into three sections (A, B, C), dependent upon the number of genital fissures and the nature of disc.

Family XIV.—OPHIURIDÆ, *Wright* (Vol. II, p. 138).

Body discoidal, depressed, having long slender rays; disc covered with hard plates, or membranous and naked. Two or four fissures (*genital*) in under side of disc in the area bounded by the rays and margin of mouth (*interbrachial space*). Rays five, always simple. Range from the Silurian to the existing period.

Section A.—Four genital fissures in each interbrachial space.

The section is represented in the British Liassic strata by one genus, Ophioderma, and five species.

Genus 30.—Ophioderma, *Müller* and *Troschel* (Vol. II, p. 140).

Disc and arms smooth, covered with small closely set granules. Arms long and slender; lateral borders provided with short papillæ or fine spines. Four genital fissures, disposed in pairs in radial lines. Range of the genus from the Liassic to existing period. Five British Liassic species (Vol. II, pp. 140—149).

Section B.—Two genital fissures in each interbrachial space. Disc provided with plates.

The section, ranging from the Trias to the existing period, is represented in the British Oolitic strata by three genera (Ophiolepis, Acroura, and Amphiura) and five species.

Genus 31.—Ophiolepis, *Müller* and *Troschel* (Vol. II, p. 150).

Disc provided with naked scales or shields on its upper surface. Two genital fissures close to the arms. Oral opening bordered with a single row of hard papillæ. Oral plates simple and heart-shaped. Lateral borders of arms bearing papillæ or spines.

Range of genus from the Trias to existing period. Three British Liassic and Oolitic species (Vol. II, pp. 150, 151, 160, 161).

Genus 32.—ACROURA, *Agassiz* (Vol. II, p. 152).

Disc small, arms long and slender, supporting on their lateral borders small scales. Range of genus from the Trias to Lias. One British Liassic species (Vol. II, p. 153).

Genus 33.—AMPHIURA, *Forbes* (Vol. II, p. 158).

Disc small, orbicular; upper surface covered with smooth small scales, the six central plates forming a rosette. Arms long, covered with scales, and exhibiting on the margins simple lanceolate spines. Range of genus from the Oolitic strata to existing period. One British Oolitic species (Vol. II, p. 158).

Section C.—*Two genital fissures on each interbrachial space. Disc membranous and naked.*

The section ranging from the Silurian strata to the existing period is represented by one British Oolitic genus (Ophiurella) and two species.

Genus 34.—OPHIURELLA, *Agassiz* (Vol. II, p. 154).

Disc small, membranous. Arms long and slender, bearing on the lateral margins elongated filiform spines. Genus found in the Oolitic strata. Two British Oolitic species (Vol. I, pp. 154, 156).

APPENDIX II.

RANGE IN GEOLOGICAL TIME

OF THE

BRITISH FOSSIL ECHINOIDEA, ASTEROIDEA, AND OPHIUROIDEA,

Described in Volumes I and II of the Oolitic Echinodermata.

	Purbeck.	Portland.	Kimmeridge Clay.	Coral Rag.	Calcareous Grit.	Oxford Clay.	Kelloway Rock.	Cornbrash.	Forest Marble.	Bradford Clay.	Great Oolite.	Stonesfield Slate.	Fuller's Earth.	Inferior Oolite.	Upper Lias.	Middle Lias.	Lower Lias.	Upper Trias.
Order I. ECHINOIDEA, *Wright*.																		
Genus 1. Cidaris, *Klein*			×	×	×					×		×?		×	×	×	×	
,, 2. Rabdocidaris, *Desor*										×				×		×		
,, 3. Diplocidaris, *Desor*														×	?			
,, 4. Hemicidaris, *Agassiz*	×	×		×			×		×	×	×	×		×				
,, 5. Pseudodiadema, *Desor*				×			×		×	×	×			×	×			
,, 6. Hemipedina, *Wright*			×	×			×			×				×	×	×	×	
,, 7. Heterocidaris, *Cotteau*														×				
,, 8. Pedina, *Agassiz*							×							×				
,, 9. Glypticus, *Agassiz*				×														
,, 10. Magnotia, *Michelin*														×				
,, 11. Polycyphus, *Agassiz*							×			×	×			×				
,, 12. Stomechinus, *Desor*				×			×			×				×				
,, 13. Acrosalenia, *Agassiz*				×			×	×	×	×	×			×			×	
,, 14. Holectypus, *Desor*				×			×			×				×				
,, 15. Pygaster, *Agassiz*				×	×		×			×				×				
,, 16. Hyboclypus, *Agassiz*										×				×				
,, 17. Galeropygus, *Cotteau*										×				×				
,, 18. Collyrites, *Deluc*			×	×	×		×							×				
,, 19. Echinobrissus, *Breynius*		×		×	×		×	×		×	×			×				
,, 20. Clypeus, *Klein*				×			×	×		×	×	×	×	×				
,, 21. Pygurus, *d'Orbigny*				×	×		×		×	×								
,, II. ASTEROIDEA, *Wright*.																		
Genus 22. Uraster, *Agassiz*										×						×		
,, 23. Tropidaster, *Forbes*																×		
,, 24. Solaster, *Forbes*											×							
,, 25. Goniaster, *Agassiz*											×			×				
,, 26. Stellaster, *Gray*														×				
,, 27. Luidia, *Forbes*																×		
,, 28. Plumaster, *Wright*																×		
,, 29. Astropecten, *Luick*				×		×	×	×			×			×		×		
,, III. OPHIUROIDEA, *Wright*.																		
Genus 30. Ophioderma, *Müller & Troschel.*																×		
,, 31. Ophiolepis, *Müller & Troschel.*														×		×	×	×
,, 32. Acroura, *Agassiz*																×		
,, 33. Amphiura, *Forbes*					×													
,, 34. Ophiurella, *Agassiz*				×				×			×							

GENERAL INDEX

TO THE

BRITISH SPECIES OF OOLITIC ECHINODERMATA

Described in Volume I (the ECHINOIDEA) *and in Volume II (the* ASTEROIDEA *and* OPHIUROIDEA).

.*.* *The names of synonyms are omitted.*

INDEX TO VOLUME II·

THE ASTEROIDEA AND OPHIUROIDEA.

26

INDEX TO THE FAMILIES, GENERA, AND SPECIES REFERRED TO IN VOLUME II.

The Synonyms are Printed in Italics.

The letters P.F. following a species denote that the fossil has been found in a Palæozoic formation.

PLATE I.

Uraster Gaveyi, *Forbes.*

From the Middle Lias.

Fig.

1 *a*. Uraster Gaveyi, *Forb.*, p. 100. Under surface, natural size.

  b. Plan of the ambulacral plates and their spiny borders.

2 *a*. Portion of the dorsal surface of a ray of *Uraster tenuispinus*, M. and T., to show the spines and pores in the tegumentary membrane.

  b. Portion of the ventral surface of the same ray, to show the biserial arrangement of the tentacula in the ambulacral valley with the lateral spines bounding the the same.

3. Diagram of the ambulacral plates in *Uraster rubens*, Lin., for the purpose of comparing them with the homologous parts in the Fossil species figured in the same plate.

Pl. I.

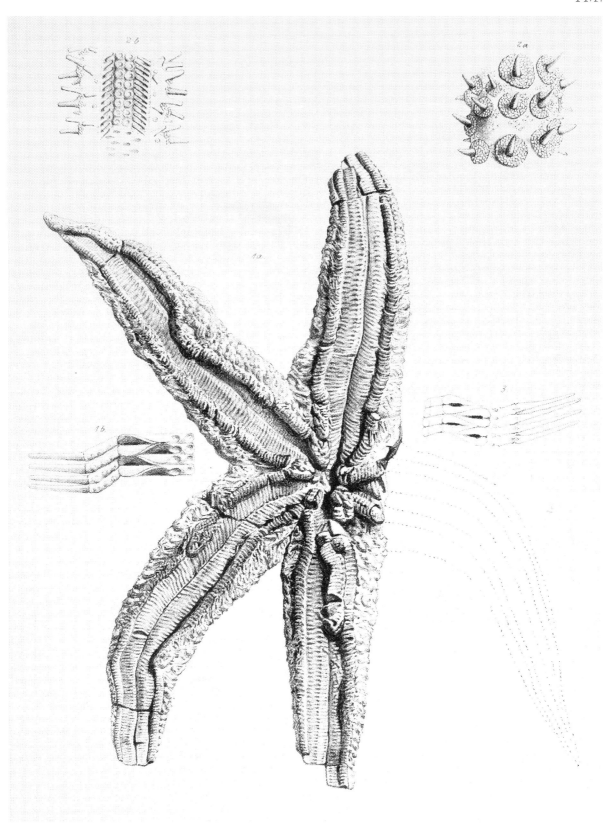

C.R.Bone del et lith.

W.West imp.

PLATE II.

URASTER CARINATUS, *Wright.*

From the Marlstone.

FIG.

1.   URASTER CARINATUS, *Wright*, p. 101.   Upper surface, natural size.

GONIASTER HAMPTONENSIS, *Wright.*

From the Great Oolite.

2 *a.*   GONIASTER HAMPTONENSIS, *Wright*, p. 109.   Upper surface, natural size.

   b.   Lateral view of the same specimen, showing the dorsal and ventral marginal plates, natural size.

From the Inferior Oolite.

GONIASTER OBTUSUS, *Wright.*

3 *a.*   GONIASTER OBTUSUS, *Wright*, p. 108.   Upper surface of a ray, natural size.

   b.   Under surface of the same ray.

   c.   Lateral view of the same, both natural size.

4 *a, b.* Marginal ossicles from an unknown Star-fish, magnified four times.

5 *a, b.*   Ditto   ditto   magnified four times.

6 *a, b.*   Ditto   ditto   magnified four times.

These ossicles were collected from the Bradford clay.

Pl. II.

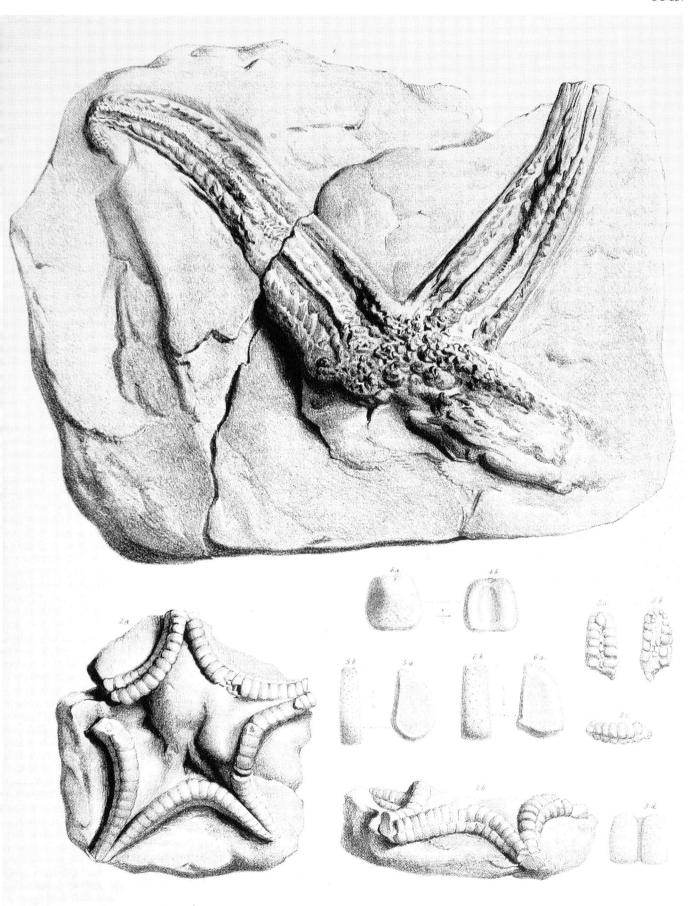

C.R.Bone del et lith.

W.West imp.

PLATE III.

Tropidaster pectinatus, *Forbes.*

From the Middle Lias.

Fig.

1. Tropidaster pectinatus, *Forb.*, p. 102. Showing the ventral surface, and a lateral view of the curved rays in small specimens.

2. Dorsal surface of two individuals on one slab, natural size.

    a. Dorsal surface of a ray magnified, to show the arrangement of the middle ridge with its imbricated plates, and the tubercles on the inter-ambulacral portions.

    b. Ventral surface of a ray, and part of the disc, showing the ambulacra and the angle-plates.

    c. Ambulacral plates, and spiniferous plates bordering the ambulacral avenue, magnified.

    d. Marginal plates, magnified.

    f. Madreporiform body, magnified.

3. A large specimen with smaller attached, both natural size.

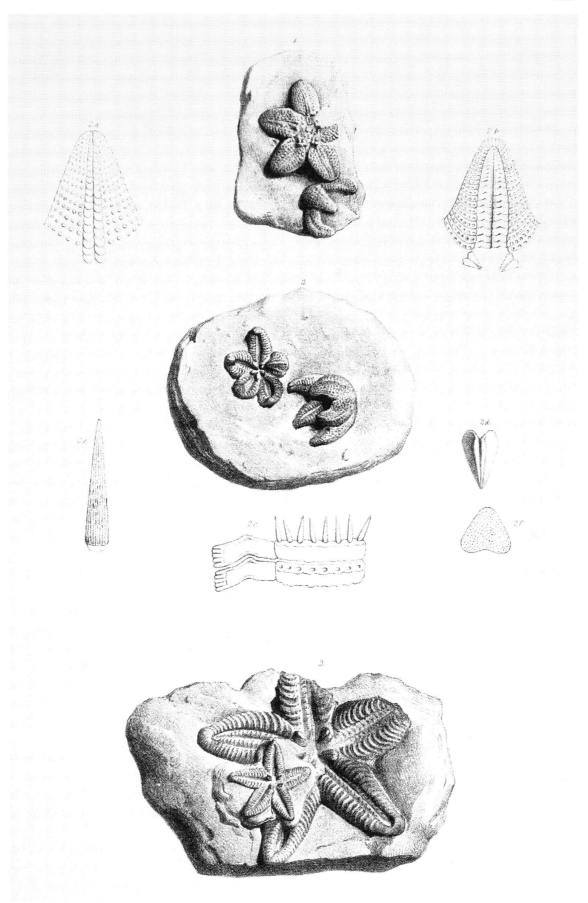

C.R.Bone del et lith.

W.West imp

PLATE IV.

SOLASTER MORETONIS, *Forbes*.

From the Great Oolite.

FIG.

1 *a*. SOLASTER MORETONIS, *Forb.*, p. 104.   Ventral surface, natural size.

 b. Ambulacral ossicles, magnified.

 c. An ambulacral bone, with inter-ambulacral ossicula supporting their combs of long hair-like spines, magnified.

 d. Inner or proximal portion on an ambulacrum with its large angle ossicula and spines, magnified.

 e. Retiform arrangement of the ossicula forming the framework of the disc, magnified.

PLATE V.

PLUMASTER OPHIUROIDES, *Wright*.

From the Middle Lias.

Fig.

1 *a*. PLUMASTER OPHIUROIDES, *Wright*, p. 112.   Ventral surface, natural size.

  b. Ambulacral and inter-ambulacral ossicula, magnified; the long inter-ambulacral plates with pectinated distal borders supporting rows of spiniferous tubercles.

  c. A pair of large angle-plates with tubercles on their surface, magnified.

LUIDIA MURCHISONI, *Williamson*.

From the Middle Lias.

2.   LUIDIA MURCHISONI, *Williamson*, p. 111.   Ventral surface, natural size.

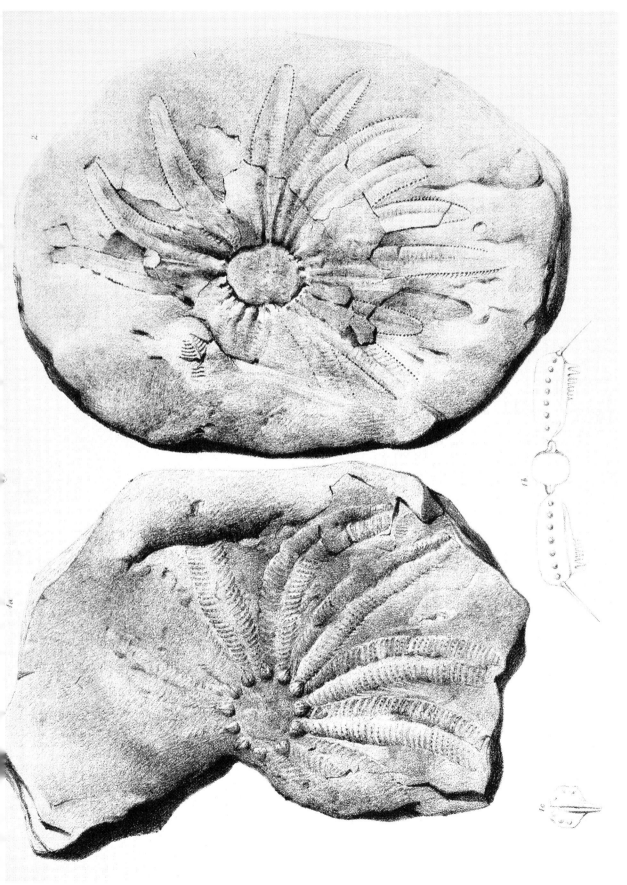

C.R. Bone del et lith.

W.West imp

PLATE VI.

Astropecten Cotteswoldiæ, var. Stamfordensis, *Wright.*

From the Stonesfield Slate.

Fig.

1 *a.* Astropecten Cotteswoldiæ, var. Stamfordensis, *Wright*, p. 118. Dorsal surface, natural size.

  b. Two marginal plates, with rows of spines at their distal border.

2 *a.* Astropecten Cotteswoldiæ, *Buck.*, p. 116. Small specimen, natural size.

  b. Four marginal plates of this specimen, showing the sculpture on their surface, the distal spines, and connecting ossicula, magnified.

Astropecten Hastingiæ, *Forbes.*

From the Marlstone.

3 *a.* Astropecten Hastingiæ, *Forb.*, p. 113. Dorsal surface, natural size.

  b. Marginal plates and inter-marginal ossicles of the same, magnified.

4 *a.* Astropecten Hastingiæ, *Forb.* Ventral surface of Mr. Leckenby's specimen, showing the ambulacral avenues, natural size.

  b. Angle-plates, ambulacral avenues, biserial pores, and inter-ambulacral plates of the same, magnified three diameters.

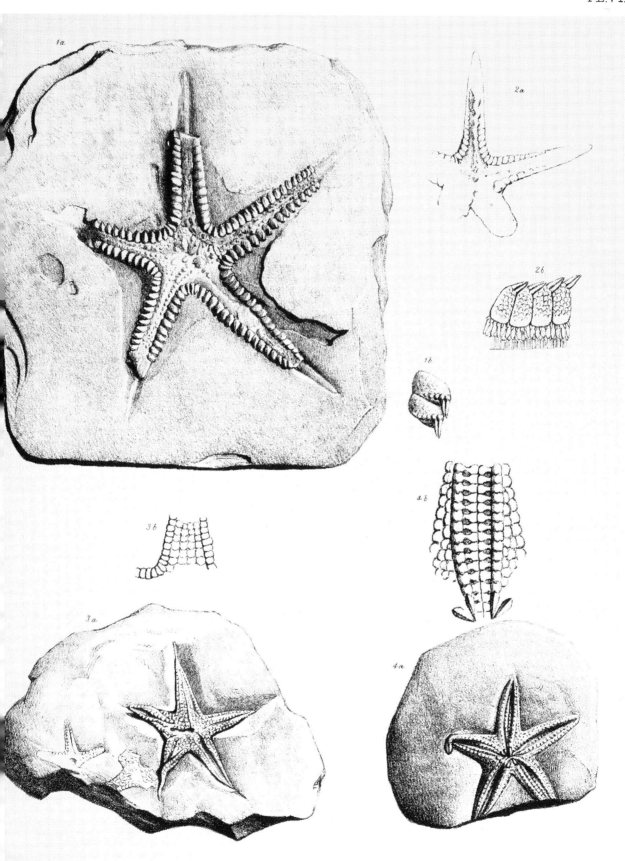

e del et lith.

W.West imp.

PLATE VII.

ASTROPECTEN LECKENBYI, *Wright.*

From the Inferior Oolite.

FIG.

1 *a.* ASTROPECTEN LECKENBYI, *Wright*, p. 114.   Dorsal surface, natural size.

  c. Two of the marginal plates, magnified four times, to show the size and arrangement of the tubercles on their surface.

  b. Lateral view of the same specimen, showing the thickness of the border and a section of the marginal plates.

ASTROPECTEN SCARBURGENSIS, *Wright.*

2 *a.* ASTROPECTEN SCARBURGENSIS, *Wright*, p. 115.   Dorsal surface, natural size.

  b. Portion of one of the rays, magnified four times.

  c. Two marginal plates, exhibiting the tubercles on their surface.

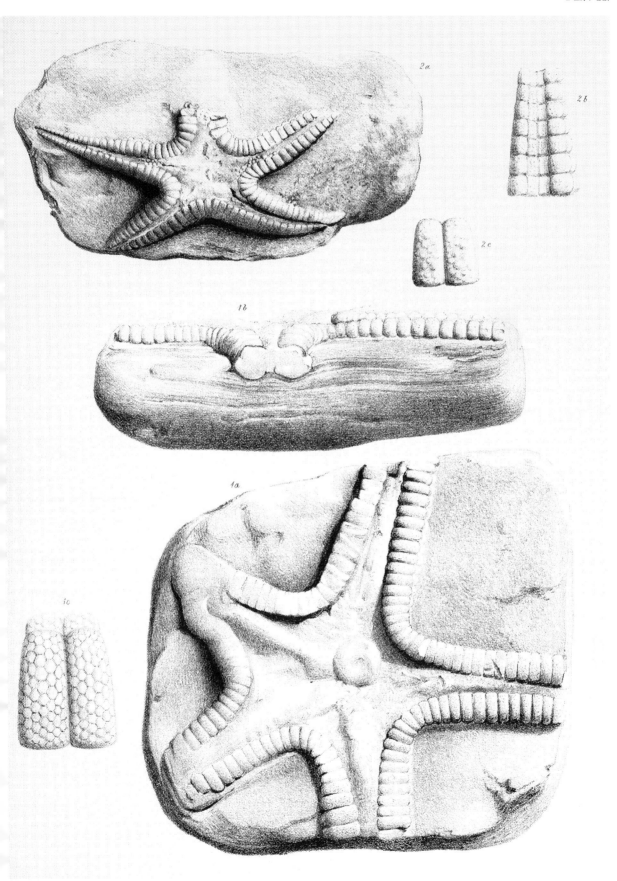

W. West imp.

PLATE VIII.

Astropecten Huxleyi, *Wright.*

From the Forest Marble.

Fig.

1 *a*. Astropecten Huxleyi, *Wright*, p. 123.   Ventral surface, natural size.

   b. Dorsal surface of the same specimen, with one ray everted, and showing both sides of the same individual at one view.

   c. Two marginal plates, with the combs of spines attached to the small inter-ambulacral bones, magnified four times.

   d. Another view of the same plates, with the large spines, magnified four times.

Astropecten Cotteswoldiæ, var. Stonesfieldensis, *Wright.*

From the Stonesfield Slate.

2.   Astropecten Cotteswoldiæ, var. Stonesfieldensis, *Wright*, p. 121.   Dorsal surface, natural size.

3.   Ossicula of Asteriadæ, magnified four times.

4.   Ditto   magnified four times.

5.   Ditto   magnified four times.

6.   Marginal plates of an *Astropecten,* magnified four times.

7.   Ditto   ditto   magnified four times.

8.   Ditto   ditto   magnified four times.

All these separate bones were collected from the Great Oolite.

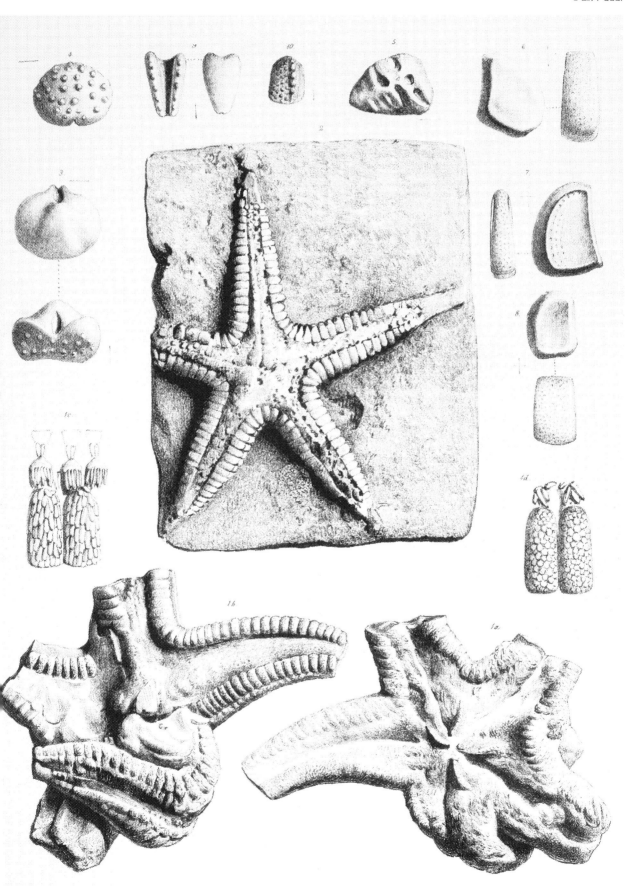

e del et lith.

W.West imp.

PLATE IX.

Astropecten Cotteswoldiæ, *Buckman.*

From the Stonesfield Slate.

Fig.

1 *a.* Astropecten Cotteswoldiæ, *Buck.*, p. 116. Dorsal surface, natural size.

  b. Disc of this specimen, magnified to show the madreporiform body with its radiating laminæ, and the five prominent oblong bodies formed by the upper portions of the ambulacral bones.

  c. Portion of the upper surface of a ray of another specimen, showing the extension of the ambulacral bones throughout the ray.

  d. Three marginal ossicula, magnified four times, showing the small granules on their convex surface.

2 *a.* Astropecten Wittsii, *Wright*, p. 120. Dorsal surface, natural size.

  b. Portion of the disc and ray, showing the madreporiform body and marginal plates of this species, magnified three times.

3 *a.* Astropecten Cotteswoldiæ. Another specimen, with a wider disc.

  b. Portion of a ray magnified twice, showing the marginal plates, and the intermarginal ossicles.

  c. Two marginal plates, magnified four times.

4. Astropecten Cotteswoldiæ. Another specimen, dorsal surface, natural size.

Pl. IX.

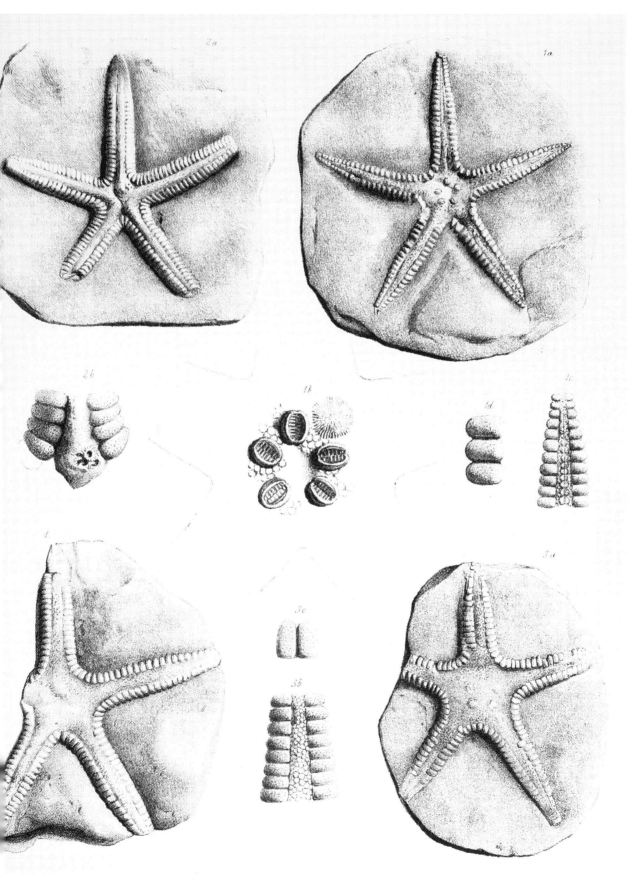

el et lith.

W.West imp.

PLATE X.

Astropecten Cotteswoldiæ, *Buck.*

From the Stonesfield Slate.

Fig.

1 *a.* Astropecten Cotteswoldiæ, *Buck.*, p. 116.   Ventral surface, natural size.

   b. Disc and portion of ray, magnified three times.

   c. Portion of the ventral surface of a ray, magnified six times, showing the marginal bones with their lateral spines, and the ambulacral avenue.

   d. Lateral view of a ray, magnified four times, showing the thorn-like spines attached to the distal border of the marginal plates.

Astropecten Phillipsii, *Forbes.*

From the Forest Marble.

2 *a.* Astropecten Phillipsii, *Forbes*, p. 122.   Ventral surface, natural size.

   b. Portion of a ray, showing the ambulacral bones, avenue, and tentacule pores, magnified.

   c. One marginal bone magnified.

   d. Two marginal bones with their border spines, magnified.

   e. Spine magnified.

Astropecten Cotteswoldiæ.

3 *a.* Astropecten Cotteswoldiæ.   Small specimen, showing details of the dorsal surface of the rays.

   b. Dorsal surface, showing marginal plates, and the upper portion of the ambulacral bones.

   c.   Ditto   ditto   of another ray.

   d.   Ditto   ditto   of another ray.

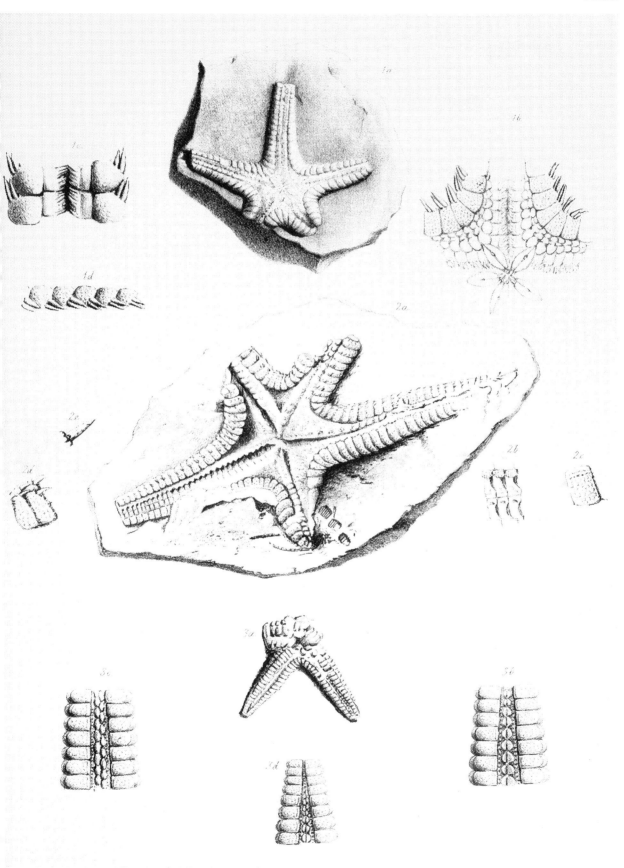

Pl. X.

Bone del et lith.

W.West imp.

PLATE Xa.

ASTROPECTEN ORION, *Forbes.*

From the Kelloway Rock.

FIG.

1. ASTROPECTEN ORION, *Forbes*, p. 127.   Ventral surface, natural size.

2. ASTROPECTEN CLAVÆFORMIS, *Wright*, p. 125.   Ventral surface, natural size; this is a four-rayed variety of the large species, figured in Plate XI.

3. ASTROPECTEN PHILLIPSII, *Forbes* (?).   Copy of the figure of an *Astropecten* found in the Cornbrash, near Yeovil, from the 'Magazine of Natural History,' vol. ii, p. 73, for 1829.

Pl. X A.

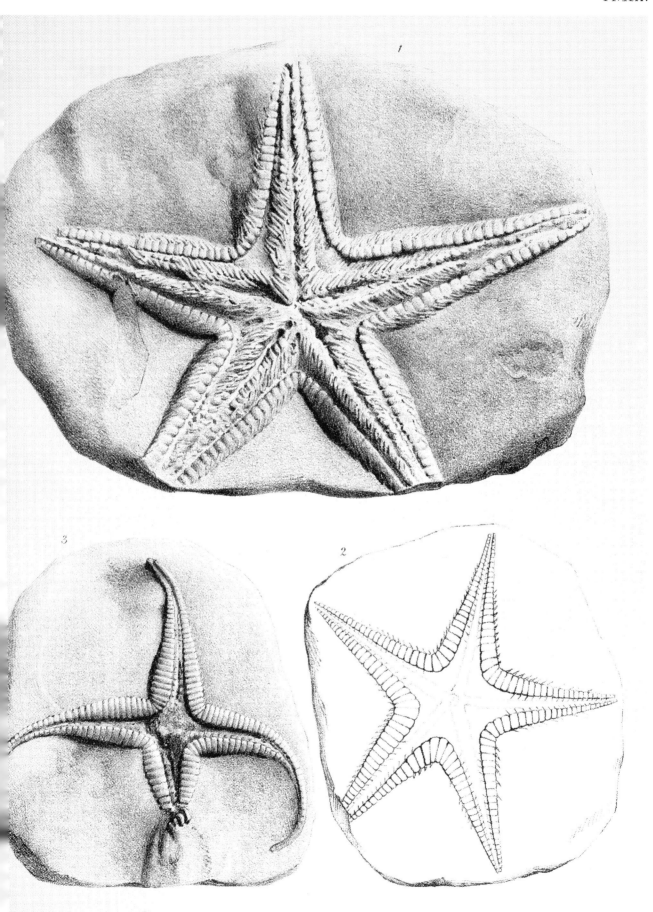

el et lith.

W.West imp.

PLATE XI.

ASTROPECTEN CLAVÆFORMIS, *Wright.*

From the Kelloway Rock.

ASTROPECTEN CLAVÆFORMIS, *Wright,* p. 125. Ventral surface, natural size.

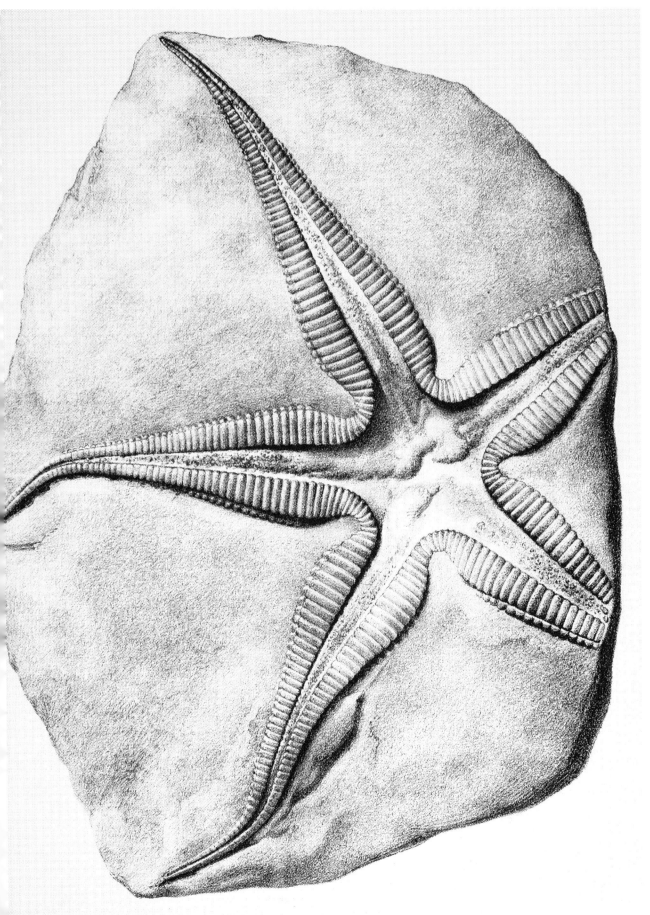

PLATE XII.

Astropecten rectus, *McCoy*.

From the Calcareous Grit.

Fig.

1. Astropecten rectus, *McCoy*, p. 129. Section of the skeleton, natural size.

2 *a.* Portion of the dorsal surface of a ray enlarged, showing the marginal plates.

   b. Portion of the ventral surface of a ray, showing the marginal plates, inter-ambulacral bones, and ambulacral avenue magnified.

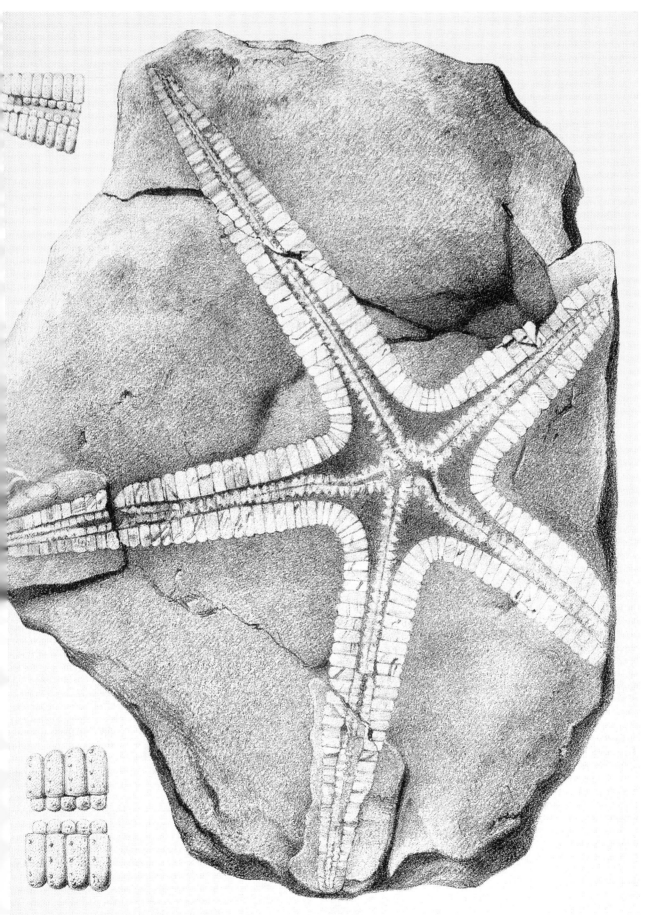

C.R. Bone del et lith.

W. West imp.

THE

PALÆONTOGRAPHICAL SOCIETY.

INSTITUTED MDCCCXLVII.

VOLUME FOR 1880.

L O N D O N :
MDCCCLXXX.

PLATE XIII.

The figures in this Plate are copied from Müller and Troschel's 'System der Asteriden.'

Pl. XIII.

PLATE XIV.

Fɪɢ.

1. Oᴘʜɪᴏʟᴇᴘɪs Mᴜʀʀᴀᴠɪɪ, *Forb.*   p. 151.
    a. Natural size.
    b. Ray, magnified.

2. Oᴘʜɪᴏʟᴇᴘɪs Mᴜʀʀᴀᴠɪɪ, *Forb.*   p. 151.   Under surface, copied from the ' London Geological Journal,' pl. **xx**, fig. 4.

3. Oᴘʜɪᴏʟᴇᴘɪs Rᴀᴍsᴀʏɪɪ, *Wright.*   p. 150.
    a. Rays, magnified.
    b. Three rings, greatly enlarged.

4. Oᴘʜɪᴜʀᴇʟʟᴀ sᴘᴇᴄɪᴏsᴀ, *Münster.*   p. 134.
5. Aᴄʀᴏᴜʀᴀ ᴘʀɪsᴄᴀ, *Münster.*   p. 133.
    a. Natural size.
    b. Portion of a ray, magnified.

6. Asᴘɪᴅᴜʀᴀ ʟᴏʀɪᴄᴀᴛᴀ, *Goldfuss.*   p. 133.
    a. Under surface.
    b. Upper surface.   Both magnified.

7. Gᴇᴏᴄᴏᴍᴀ Lɪʙᴀɴᴏᴛɪᴄᴀ, *König.*   p. 134.   Under surface, magnified.
8. Aᴘʟᴏᴄᴏᴍᴀ Aɢᴀssɪᴢɪɪ, *Münster.*   p. 134.   Under surface, magnified.
9. Pʀᴏᴛᴀsᴛᴇʀ Sᴇᴅɢᴡɪᴄᴋɪɪ, *Forbes.*   p. 136.   Under surface, magnified.
10. Aᴍᴘʜɪᴜʀᴀ ᴛᴇɴᴇʀᴀ, *Lütken.*   p. 136.
    a. Under ⎫ surface of disk, magnified.
    b. Upper ⎭

Pl XIV.

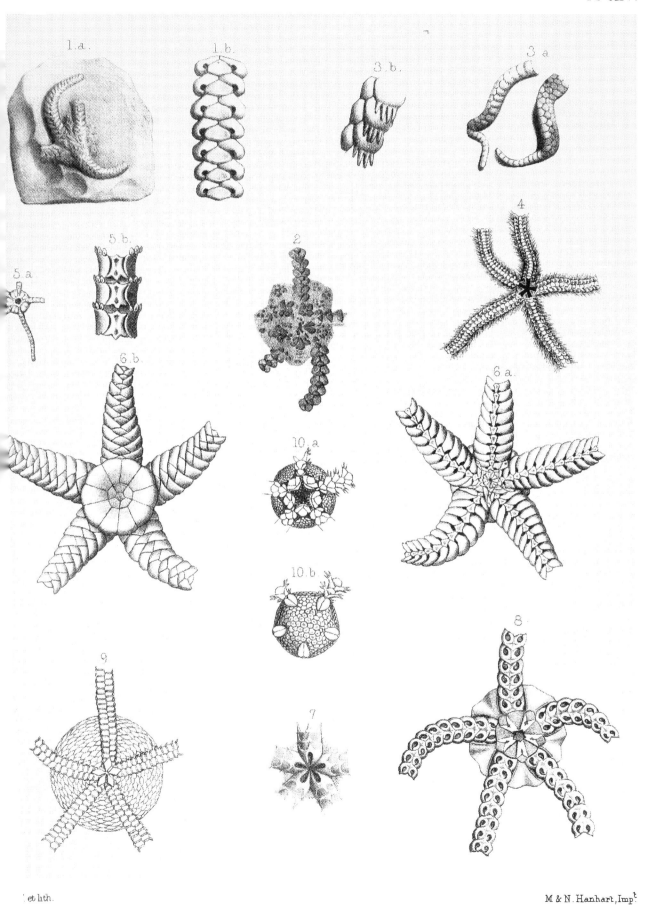

et lith.

M & N. Hanhart, Imp.t

PLATE XV.

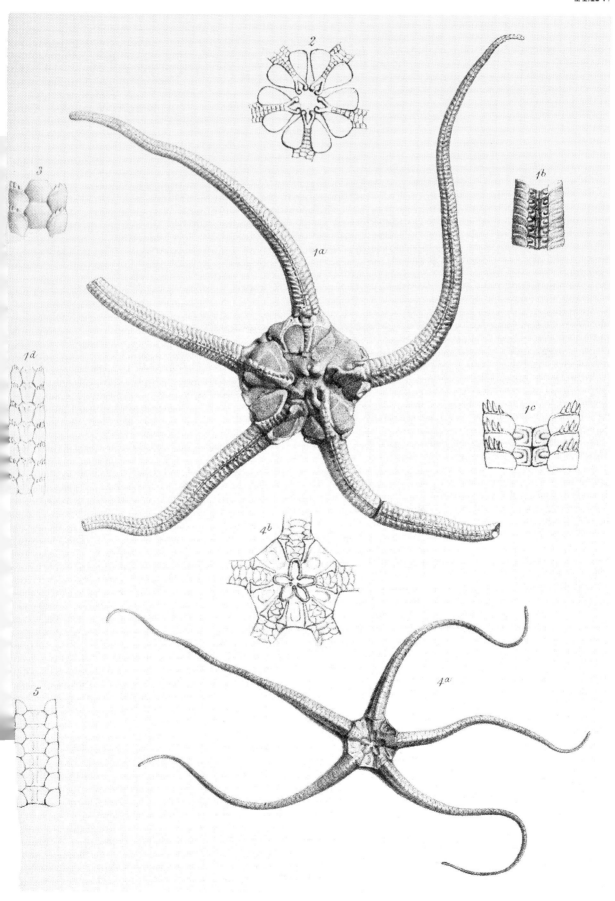

. del et lith.

W. West imp.

PLATE XVI.

Pl. XVI.

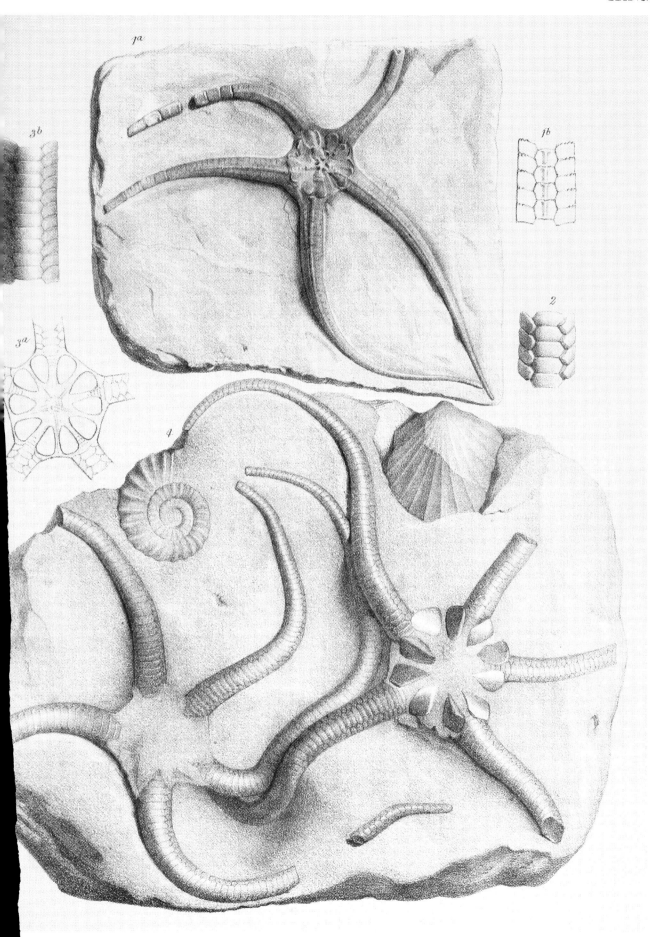

PLATE XVII.

Pl. XVII.

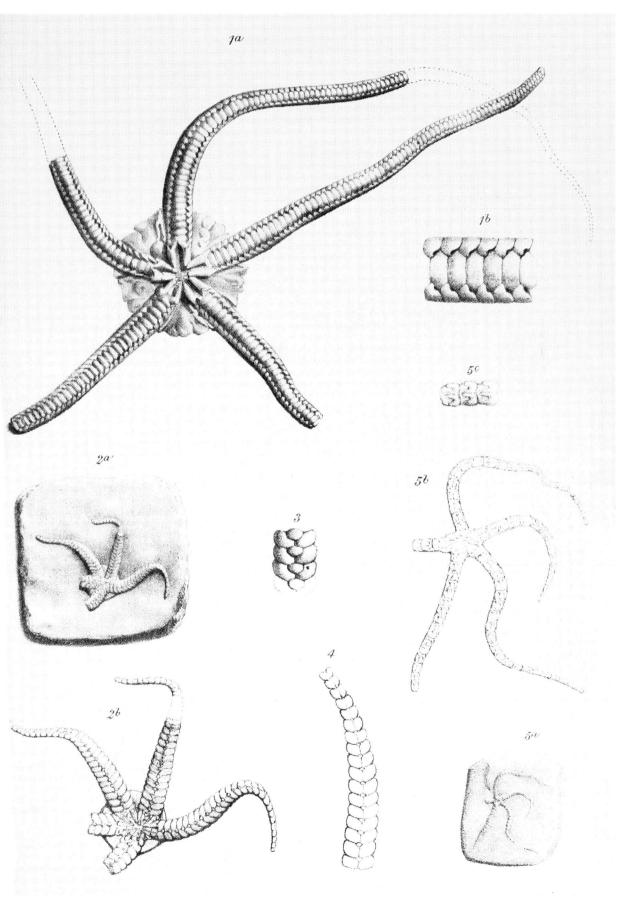

PLATE XVIII.

Pl. XVIII.

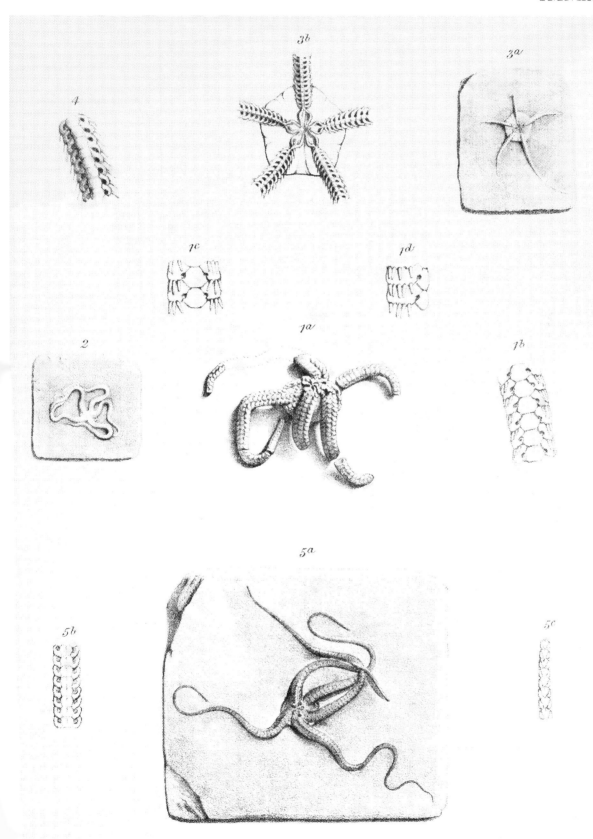

R. Bone del et lith.

W. West imp.

PLATE XIX.

Fig. 1 *a*. ASTROPECTEN RECTUS, *McCoy*. Showing the form and structure of the marginal ossicles, a rare occurrence in this species. Woodwardian Museum, Cambridge. Leckenby collection.

1 *b*. — — — Three marginal plates enlarged, showing the socket-like depressions on each alternate plate, and the cellular structure in all.

2. OPHIOLEPIS MURRAVII, *Forbes*. Marlstone near Staithes, enlarged two diameters. My collection.

3 *a*. — LECKENBYI, *Wright*. Upper surface, natural size. My collection.

3 *b*. — — — Three ring elements from the upper segment, magnified.

4. — — — Under side of the disk, showing the mouth and origin of the rays magnified.

Pl. XIX.

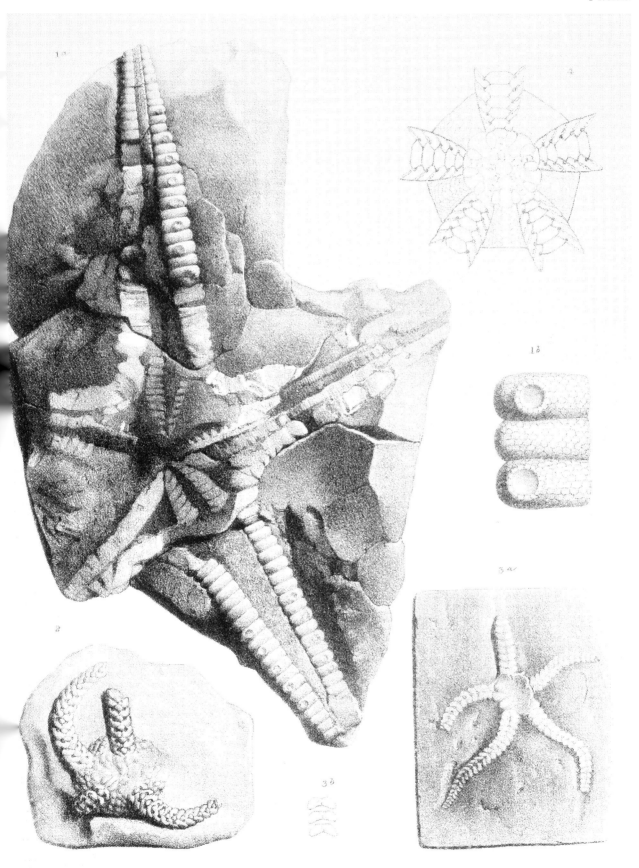

M & N. Hanhart, Imp.t

STELLASTER SHARPII, *Wright.*

From the Inferior Oolite.

This grand Jurassic *Goniaster* was obtained from the Inferior Oolite near Northampton by my old esteemed friend Samuel Sharp, Esq., F.G.S., in whose collection it was contained when I figured the specimen. It now forms part of the Jurassic Echinodermata collection from the Inferior Oolite of Northamptonshire in the British Museum.

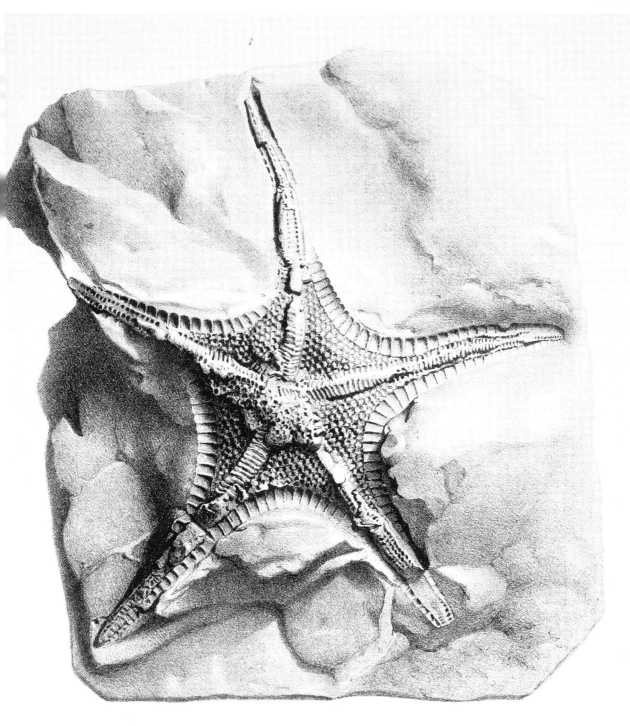

del et lith.

M & N. Hanhart, Imp^t

PLATE XXI.

Fig. 1. URASTER SPINIGER, *Wright*.   Under surface (reversed by the artist), magnified two diameters, from the Forest Marble near Road, Wiltshire.   My collection.

2. STELLASTER BERTHANDI, *Wright*.   Under surface, natural size.   This beautiful form was collected from the " Calcaire à Entroques," Inferior Oolite, near Mâcon, by Professor Berthandi, of Mâcon, Saone-et-Lôire, France.   My collection.

3. ASTROPECTEN HOOPERI, *Wright*.   The under surface of this Astropecten, which was collected from the Cornbrash at Horsington, Dorset, in 1828.   The figure is a copy of a woodcut published in Loudon's ' Magazine,' vol. ii, p. 73. The specimen cannot be traced, and therefore the original drawing is reproduced for this work.

4. OPHIOLEPIS DAMESII, *Wright*.   Under surface, magnified two diameters.   My collection.

5.   ——   ——   ——   Upper surface of another specimen, magnified two diameters.   Collected from the *Avicula contorta* beds at Garden Cliff, near Westbury, on the Severn.   The first known ginal specimens were found in the *Avicula contorta* bed near Hildesheim, N. Germany, and the same species has since been found in the same horizon in England at two or three localities.

Pl. XXI.

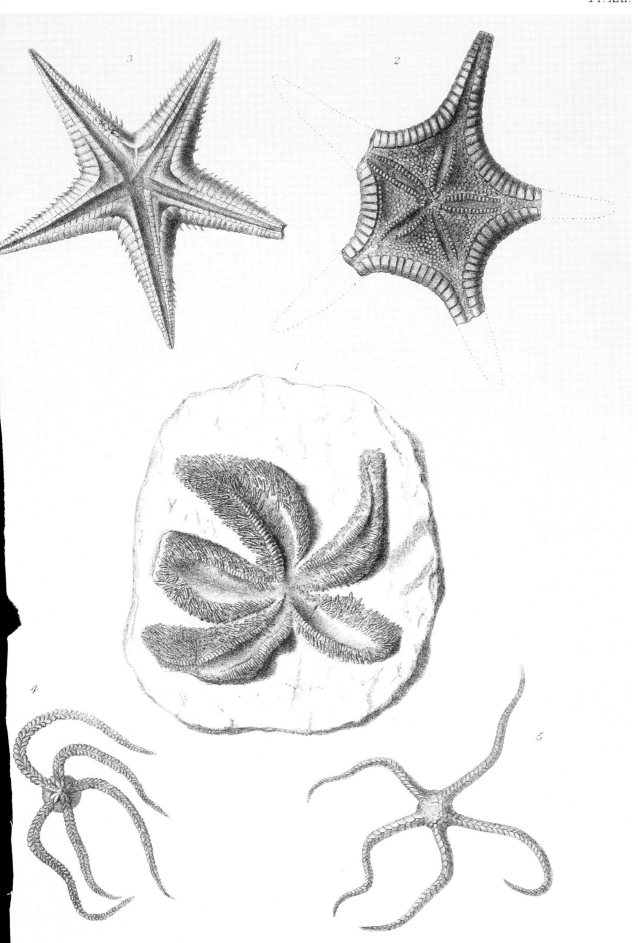

. et lith.

M & N. Hanhart, Imp.ᵗ

Printed in the United States
By Bookmasters